A Practical Guide to the 17th Edition of the Wiring Regulations

A Practical Guide to the 17th Edition of the Wiring Regulations

Christopher Kitcher

AMSTERDAM • BOSTON • HEIDELBERG • LONDON
NEW YORK • OXFORD • PARIS • SAN DIEGO
SAN FRANCISCO • SINGAPORE • SYDNEY • TOKYO
Newnes is an imprint of Elsevier

Newnes is an imprint of Elsevier
The Boulevard, Langford Lane, Oxford OX5, 1GB, UK
30 Corporate Drive, Suite 400, Burlington, MA 01803

First edition 2010

Notices
Knowledge and best practice in this field are constantly changing. As new research and experience broaden our understanding, changes in research methods, professional practices, or medical treatment may become necessary.

Practitioners and researchers must always rely on their own experience and knowledge in evaluating and using any information, methods, compounds, or experiments described herein. In using such information or methods they should be mindful of their own safety and the safety of others, including parties for whom they have a professional responsibility.

To the fullest extent of the law, neither the Publisher nor the authors, contributors, or editors, assume any liability for any injury and/or damage to persons or property as a matter of products liability, negligence or otherwise, or from any use or operation of any methods, products, instructions, or ideas contained in the material herein.

British Library Cataloguing-in-Publication Data
A catalogue record for this book is available from the British Library

Library of Congress Cataloging-in-Publication Data
A catalog record for this book is available from the Library of Congress

ISBN: 978-0-08-096560-4

For information on all Newnes publications
visit our website at www.newnespress.com

Typeset by MPS Limited, a Macmillan Company, Chennai, India
www.macmillansolutions.com

Printed and bound in Italy

10 11 12 13 14 10 9 8 7 6 5 4 3 2 1

Contents

Part 5

Part 6

Appendices

Foreword

Welcome to this book on the Wiring Regulations. I am proud to have known Chris as both a friend and work colleague for the last 15 years, delivering Electrical Installation, both theory and practice, to students at Central Sussex College. What is obvious about Chris is not only his enthusiasm for the subject but also the dedication he has to passing his knowledge on in a way that can be understood. None of us is likely to be a rocket scientist, but sometimes to read the Wiring Regulations you might think that was who they were written for. Phrases such as 'prevention of mutual detrimental influences' make you wonder just what you have let yourself in for. But of course the Wiring Regulations have to be concise as after all they are a set of rules. What Chris has achieved in this book is a clear, concise interpretation of those rules to assist your understanding of the Wiring Regulations. His experience as a lecturer in putting this knowledge across clearly shows through the explanations and clear diagrams throughout the book.

Richard Brooks Cert Ed LCGI MIET
Director of Technology – Central Sussex College

Preface

This book is not intended to take the place of the 17th Edition of the Wiring Regulations (BS 7671); instead I hope that it will be used as a reference book alongside BS 7671.

I have been involved in the construction industry all of my working life, primarily in the electrical side. When I left school and started my first work, I was 15 years old, at which time I was quite sure that I would never attend another lesson in my life. To me, school was an absolute waste of time and I could not face the thought of spending another day sitting behind a desk. I am pretty sure that I was not the only person to feel like that, as I was more of a practical person and not at all academic.

Back in the early 1960s, there was less emphasis on gaining a qualification than there is now, although of course it was desirable. Gaining a qualification would have required attending a technical college; so given my view of education, there was to be no qualification for me! I took a job as an electrician's labourer working for a small contracting firm which was involved in all types of electrical contracting and repair.

I was incredibly lucky as the firm was run by two lovely brothers, Tony and Ron Pointing, who took me under their wing, showed an unbelievable amount of patience and gave me an incredible apprenticeship. This book is dedicated to them, as without them, I would probably still be a labourer.

Twenty-five years after I left school I decided that it may be a good idea to attend an electrical course at Crawley College to see if I could gain an electrical qualification. The practical side of the course was very easy for me, as it was what I had been doing for the last 25 years. Unfortunately, the theory side of the course was a completely different kettle of fish, and I could not begin to put into words how difficult I found it all. To help me with my studies, I bought countless books which I read from cover to cover, generally several times. Unfortunately, many of these books were still too complex for me to fully understand, but I persevered and gained the required qualifications. I remember thinking at the time that if I found the course difficult, what must it be like for younger students who had virtually no practical experience?

After I qualified, I was asked to do a bit of part-time teaching at a college and was initially a bit apprehensive about it but decided to give it a try. Right from the first day, everything seemed to go pretty well and I have been teaching ever since. I very quickly learnt that many students who wanted to become electricians were just like me and found the academic side of the course very difficult. Because of this, I took a different approach and started writing lessons in my own words instead of using the usual text books, which although they were very good, did not really suit my way of learning.

Some time ago, when I wrote a book called *A Practical Guide to Inspection, Testing and Certification of Electrical Installations*, I tried to keep it as simple as possible and to write it in a way which I would have liked to have been taught. This appeared to work as the book seems to be quite popular, and for that reason when I undertook to write this book, I decided to write it in the same style. I must admit that I found it quite a challenge but I hope that readers will find the contents at least useful, if not interesting.

I would like to acknowledge the help, information and facilities provided to me for the writing of this book by Central Sussex College, Megger UK for allowing the use of their drawings and test equipment, the NICEIC for their assistance and for allowing the reproduction of their certificates, Hager for allowing me to use information from their technical guide, Bender UK Ltd for providing information on monitoring equipment, and my colleagues, Jonathan Knight and Gary Maunder, for proof reading all of the countless documents which I kept providing for them, without complaint.

Last but not least, I would like to thank my very understanding wife Jill for not insisting that I went shopping with her when I had this book to finish.

BS 7671 Wiring Regulations 2008

BS 7671:2008 is the British Standard for electrical installations in the United Kingdom; it is a non-statutory document. The statutory document with which persons installing or working on electrical installations must conform is the Electricity at Work Regulations 1989 (EAWR 1989).

The EAWR 1989 are one of the many sets of regulations which fall under the umbrella of the Health and Safety at Work Act 1974 (HASAWA 1974). Compliance with BS 7671:2008 will almost certainly mean that the requirements of the Health and Safety Executive will be satisfied.

The purpose of the Wiring Regulations is to protect persons, property and livestock from harm which could arise from the use or presence of electricity.

BS 7671:2008 should be referred to for all new installations, and any additions and alterations to existing installations; it must be remembered that installations which were installed using earlier editions of BS 7671 will not become non-compliant unless alterations or additions are carried out on them. When an existing installation has a circuit added to or altered the circuit must then be compliant with the latest edition of BS 7671. In some cases, this will require the completion of an electrical installation certificate even if the alteration is simply adding to an existing circuit (*this will be explained in more detail later*).

As BS 7671:2008 is a non-statutory document other methods can be used to gain compliance with EAWR 1989; however, it is probably better to comply with the latest regulations as in most cases they will be found to cover any given situation, and the installer will be protected in law.

A Practical Guide to the 17th Edition of the Wiring Regulations. DOI: 10.1016/B978-0-08-096560-4.00011-4

Part 1

In this Chapter:

Part 1 describes the range, purpose and essential principles required for compliance with BS 7671. It also identifies the type of installations to which it should be applied, along with any type of installations which are not included.

Regulation 101.1 provides a list of installations which are included in the scope of BS 7671. They are:

- Residential, commercial, industrial and agricultural/horticultural premises
- Caravans, caravan parks and prefabricated buildings
- Construction and demolition sites, fairgrounds, exhibition and shows and any other temporary electrical installations
- Marinas
- External lighting, highway equipment and street furniture
- Mobile and transportable units
- Photovoltaic systems and low-voltage generating sets.

Whilst working on any electrical installation, it should be remembered that other British Standards (BS) must be consulted where they may have an effect on the installation. These installations could include:

- BS 559 and BS EN 50107 for electric signs and high-voltage luminous discharge tube installations
- BS 5266 for emergency lighting
- BS EN 60079 for electrical equipment for explosive gas atmospheres
- BS EN 50281 and BS EN 61241 for electrical equipment for use in the presence of combustible dust
- BS 5839 for fire detection alarm systems in buildings
- BS 6701 for telecommunication systems
- BS 6351 for electric surface heating systems

A Practical Guide to the 17th Edition of the Wiring Regulations. DOI: 10.1016/B978-0-08-096560-4.00001-1

- BS 6907 for electrical installations for open-cast mines and quarries
- BS 7909 for design and installation of temporary distribution systems delivering a.c. electrical supplies for lighting, technical services and other entertainment-related purposes.

This is not an exhaustive list, and if in doubt when working on an installation, enquiries about the use of other standards should be made.

LICENSING LAWS

Some premises are subject to licensing laws; in other words the local authority may have certain requirements that need to be met before the owners of the installation are granted a licence to operate. These installations include petrol stations, caravan sites, etc. It should be remembered that any additional requirements needed by these statutory bodies are in addition to the requirements set out in BS 7671, not instead of them.

All equipment and materials used in an electrical installation must be of a British Standard (BS) or European norm (BS EN); any equipment or materials which comply with these standards will also comply with BS 7671 provided that they are used for the purpose for which they are intended. It must be remembered that some equipment is only satisfactory when used in a certain way or with suitably matching equipment.

MANUFACTURER'S INSTRUCTIONS

It is a requirement of BS 7671 that all electrical equipment is installed in accordance with the manufacturer's instructions and documentation. Where this information is not available or has been lost, a call to the manufacturer or supplier will usually provide the required information.

BRITISH STANDARDS

From time to time a designer/electrician may be asked to install equipment which has no standard marked on it; this could possibly be a new invention or new type of material.

Regulations 102.4, 103.1.2 and 103.1.3 cover this situation, and make it very clear that they can be used provided the equipment or materials in question provide the same degree of safety required by BS 7671. This must be agreed between the designer or client and the person responsible for the installation. In many instances, the designer and installer could be the same person. It is not unusual for electricians to be expected to design and install an installation, with the customer often asking that we install equipment supplied by them.

If we are confident that on completion the installation will provide the same degree of safety as expected by BS 7671, it is permissible to use the equipment in the installation. It is also a requirement that we record the fact that we have departed from the requirements of BS 7671 by making a note of the departure on the electrical installation certificate.

It can be seen from the electrical installation certificate in appendix 6 that the designer, installer and the person responsible for carrying out the inspection and test must all agree that the departure is acceptable. If agreement cannot be met then the departure should not be allowed.

CHAPTER 13

Fundamental principles

Where BS 7671 is complied with, a high level of safety protection is provided for any persons, property and livestock which come into contact with the installation.

A good working knowledge of the contents of chapter 13 is vital for any person who is going to design an electrical installation. This chapter sets out the basic principles which are required to be considered within an installation.

Before we can consider designing an installation, a full understanding of these basic principles is very important. Most electricians will know what the fundamental principles are and take most of them into account without a second thought. If these fundamental principles are read carefully, it will be seen that most of them are just common sense requirements.

However, I am pretty sure that most of us would struggle to make a complete list of them. Over the next few pages each fundamental principle is listed and briefly explained. Each principle will also be cross-referenced throughout this book.

131 Protection for safety

The whole reason for BS 7671 being in place is to provide a standard which can be used to ensure the safety of persons, property and livestock when electrical installations are used in the way in which they are designed to be used. BS 7671 also provides a level of safety within installations when things go wrong, during moments of misuse or when accidents occur.

Regulation 131.1 provides us with a general list of possible occurrences which we must consider providing protection against where required. These are:

- Shock currents
- Excessive temperatures
- Explosion
- Under- and overvoltages along with electromagnetic influences
- Mechanical movement of electrically operated equipment
- Arcing or burning
- Power supply interruptions.

Any of the items listed could cause severe injury or damage depending on the circumstances.

Sections 131 to 135 expand on the general list and tell us what we need to consider for each item.

131.2.1 Basic protection

The installation must have in place a method of preventing persons or livestock from touching live parts of the installation which are intended to be live where any contact could result in a current passing through the body, or to limit the current which could pass through a body to a safe level.

Methods of protection are set out in part 4 of the Wiring Regulations. They are:

- Basic insulation of live parts
- Barriers or enclosures
- Obstacles and placing out of reach
- Use of separated extra low voltage (SELV) or protected extra low voltage (PELV).

Basic protection can be achieved by complying with 414, 416 and 417 of BS 7671.

131.2.2 Fault protection

Fault protection is required to protect persons and livestock against electric shock which may arise from contact with exposed or extraneous conductive parts which may become live during a fault.

This can be achieved by any one of the following methods:

- Preventing a current passing through the body of persons or livestock
- Limiting the duration of the fault current that could pass through a body by the fast operation of a protective device
- Limiting the level of current that could pass through a body in the event of a fault.

Correct use of protective devices along with protective bonding is the most commonly used measure for fault protection as it provides automatic disconnection of supply; this is referred to as ADS. Other recognised methods of fault protection are the use of double or reinforced insulation or SELV or PELV. (Regulation 410.3.3.)

131.1 Protection against thermal effects

Where there is electricity there is a very good chance that there will be heat. The cause of this heat could be:

- Current flowing in conductors
- Heat radiated by lamps, heaters or motors
- Loose connections
- Arcing.

An electrical installation must be protected against any harmful effects that could be caused due to this heat.

Too much current flowing in conductors could easily cause a fire, as could loose connections. Loose connections could also cause serious damage to equipment due to arcing and overheating.

Overheating could cause damage to the surfaces to which the equipment is fixed, as well as cause flammable materials to ignite.

Radiated heat could cause serious damage to surfaces, and possibly fire, as well as cause injury to persons and livestock.

131.4 Protection against overcurrent

Overcurrent in a circuit or installation can cause damage due to high temperatures being generated, and in the case of very high currents, electromechanical stresses can occur.

The correct selection of conductor sizes, protective devices and equipment will go a long way to ensuring that overcurrent does not create a problem.

131.5 Protection against fault current

Protection against fault current in live conductors will be achieved by ensuring the requirements of 131.5 are satisfied, in particular the selection of the correct type of circuit breakers or fuses. Bonding to other parts of the installation that will be required to carry fault currents must be selected carefully. This is to ensure that excessive temperatures or physical damage do not occur.

In general terms, the use of equipment to British Standards will satisfy this requirement.

131.6 Protection against voltage disturbances and measures against electromagnetic disturbances

Undervoltage or loss of supply would be the most common occurrence in most installations. Systems must be in place to prevent injury due to the sudden starting of machinery when a supply is reinstated.

Overvoltages due to switching or lightning strikes along with electromagnetic disturbances must also be considered. The use of British Standard equipment and in some cases surge protection will generally be enough to satisfy this requirement.

Where there is a possibility of lightning strikes BS EN 62305 should be referenced.

131.7 Protection against power supply interruption

If danger or damage could arise due to the loss of the supply suitable precautions must be taken, which could be as simple as providing emergency lighting if required. Where the damage could be to equipment or life support machinery then a backup supply would need to be considered and chapter 56 of BS 7671 must be complied with.

131.8 Additions and alterations to an installation

Before any **additions** or **alterations** are made to an electrical installation, the person carrying out the work must make sure that the installation is suitable to add or alter and that it will remain safe after the alterations have been completed. On completion of any work, the earthing and bonding arrangements must comply with the latest edition of the Wiring Regulations.

132.1 Design

Clearly, all installations have to be designed to do what they are intended to do as well as providing the best measures of protection possible for persons, livestock and property. To enable us to design an installation or even a single circuit correctly, we have to have information available on the following.

132.2 Characteristics of supply

- Voltage and frequency
- Number of conductors
- Type of supply TT, TN-S, TN-C-S
- Maximum current which can be drawn from the supply
- Z_e
- Prospective short-circuit current
- Any requirement specific to the supplier.

132.3 Nature of demand

This really comes down to how much current will be required for the safe operation of the installation allowing for any future reasonable additions. Any special conditions of the installation must be taken into account. Diversity can also be used where required and permissible.

132.4 Electrical supply systems for safety services or standby electrical supply systems

If a safety service supply is required, information must be available on the characteristics as in 132.2 and the circuits which are to be supplied. In a large installation, it would be unusual for a safety service to be made available for the whole installation.

132.5 Environmental conditions

All installed equipment must be suitable for any conditions which may affect it. This would include weather, dust, heat, vibration and impact. In some installations, vapour-proof and fireproof equipment may be required.

Having ascertained the information for the basis of the design from the information required in regulations 132.2 to 132.5 the installation must now be designed to comply with regulations 132.6 to 132.15.

132.6 Cross-sectional areas of conductors

The size of conductors must be calculated to withstand the current which is required to be carried under normal operating and fault conditions without damage. Unfortunately, it is not just a matter of looking at a table and referencing the type of cable and the current it can carry. Cable selection can be quite complex, as we will see later in the book.

When selecting a cable, consideration must be given to:

- Current carrying capacity required
- Permissible voltage drop
- Maximum operating temperature
- Method of installation
- Mechanical stresses due to environmental conditions
- $R_1 + R_n$
- $R_1 + R_2$
- Electromechanical stresses which could occur in the event of a short circuit or an earth fault
- Thermal insulation which could affect the current carrying capacity of the conductor
- Harmonics.

132.7 Types of wiring and method of installation

When choosing the type of wiring system and the method of installation, it is very important that consideration is given to the environment, utilisation and construction of the building as stated in appendix 5. The wiring system which is selected must be unaffected by any situations which may be present due to these external influences. Accessibility, maintainability and electromagnetic interference must also be considered.

132.8 Protective equipment

All protective equipment installed must be suitable for any overcurrent, fault current, overvoltage, undervoltage and no-voltage. Consideration must also be given to voltage and disconnection times.

132.9 Emergency control

Where required, emergency stop buttons or switches must be installed to comply with regulation 537.4.

132.10 Disconnecting devices

It must be possible to isolate circuits or equipment for inspection, maintenance, testing and operation. Any devices must comply with chapter 53 of BS 7671.

132.11 Prevention of mutual detrimental influence

Consideration must be given to any damage or interference that could occur due to the wiring system being placed too close to any non-electrical equipment; this could be due to heat transference, corrosion or the necessity of other trades having to carry out work on other services. Correct segregation of band I and II circuits or safety circuits must also be considered.

132.12 Accessibility of electrical equipment

All electrical equipment and accessories must be accessible for maintenance, inspection and repair when required. This includes the complete removal of any equipment covers or enclosures. It is permissible to install junction boxes under floors, etc.; however, it is important that they can be accessed. Suitable access traps should be left along with information indicating where the traps can be found; the access point can be covered by floor covering which may have to be removed for access.

132.13 Documentation for the electrical installation

Documentation must be provided for each installation, or addition to the installation.

For a new installation or circuit, an electrical installation certificate accompanied by a schedule of test results and a schedule of inspections would be the minimum requirement.

For an addition to an existing circuit, a minor electrical installation works certificate is required, provided the characteristics of the circuit remain unaltered. In some cases, an electrical installation certificate, accompanied by a schedule of test results and a schedule of inspection, would be required, depending on the extent of the work carried out. An example of this would be where a protective device was changed for a different type.

132.14 Protective devices and switches

Single pole switches and protective devices must only be installed in the line conductor, where it is required to switch a neutral as well as the line conductor. Double pole devices must be used.

In the supply systems which are in general use in the UK, the neutral will always be effectively connected to earth. While there is no regulation which prevents us from switching the neutral (unless it is a protective earth and neutral (PEN) conductor), an earthed neutral must never be switched unless the switch is linked and disconnects the line conductors of the circuit at the same time as the neutral.

132.15 Isolation and switching

This is really the same requirement as 132.1 except it mentions electric motors.

133 Selection of electrical equipment

This section covers the choice of electrical equipment. Any equipment used must be suitable for use with the voltage, current, frequency, power and installation methods used.

As referred to earlier, use of equipment to a suitable British Standard will ensure that this part of BS 7671 is complied with. In the absence of a British Standard, agreement must be reached between the person who has specified the equipment and the person responsible for the installation, and a record of this must be made in the appropriate section of the electrical installation certificate.

Where equipment is used which has no British Standard it must offer at least the same degree of safety that would be required by BS 7671.

134 Erection and verification of electrical installations

134.1 Erection

Provided the installation is installed to comply with BS 7671 compliance with this and all other parts of the regulations will be achieved.

134.2 Initial verification

From the moment work on an electrical installation commences, inspecting and testing must be carried out. At the beginning of the work this would be simply a visual check on materials as they are installed. This would increase as the work progresses with some testing being carried out where required, particularly where long gaps between visits to the site occur, and elements of the installation have been covered up by the work being carried out by other trades.

On completion of the installation, a final inspection and test is to be carried out (initial verification) by competent persons to ensure that the installation is to the standard required by BS 7671. On completion of the initial verification, the appropriate documentation must be completed and signed as required. The designer must recommend the interval to the first periodic test after taking into account the use, environment and type of installation which is being certificated.

135 Periodic inspecting and testing

All installations must be inspected and tested periodically to ensure that they are safe for continued use. As stated earlier, the designer will decide when the first inspection should be carried out. The interval between further inspections will be decided by the person carrying out the inspection.

Compliance with chapter 13 will ensure that all of these principles have been addressed and that the best possible protection is in place.

Part 2

In this Chapter:

DEFINITIONS

This section of BS 7671 consists of definitions and is very useful for reference, particularly in the electrical industry where we use our own form of 'slang'. Of course, we know what we mean but occasionally we have to write reports, perhaps for surveys and periodic inspections. It is very important that these reports are written correctly and that the correct terminology is used to reduce the risk of any misinterpretation.

For example, there is often confusion with the identification of earthing and protective conductors (Figure 2.1).

It is also important to be able to recognise the different types of supply systems which are used in the UK. The systems which we will normally be required to work on are TT, TN-S and TN-C-S.

The letters are:

T for terra, which is the Latin word for earth
N for neutral
C for combined
S for separate.

TT SYSTEM

Overhead systems are still very common. It was far cheaper and quicker to get electricity into homes by running cables above ground than to bury them. When these systems were installed they were TT systems. They can be easily identified by looking at the service head and the supply cable entering into it.

A Practical Guide to the 17th Edition of the Wiring Regulations. DOI: 10.1016/B978-0-08-096560-4.00002-3

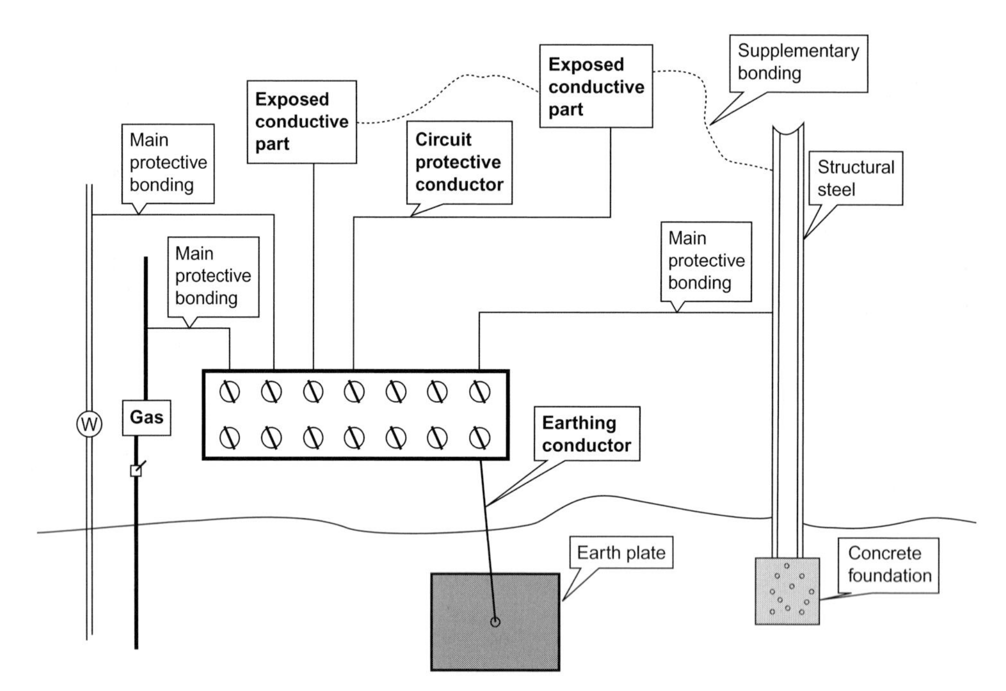

FIGURE 2.1 **Protective conductors**

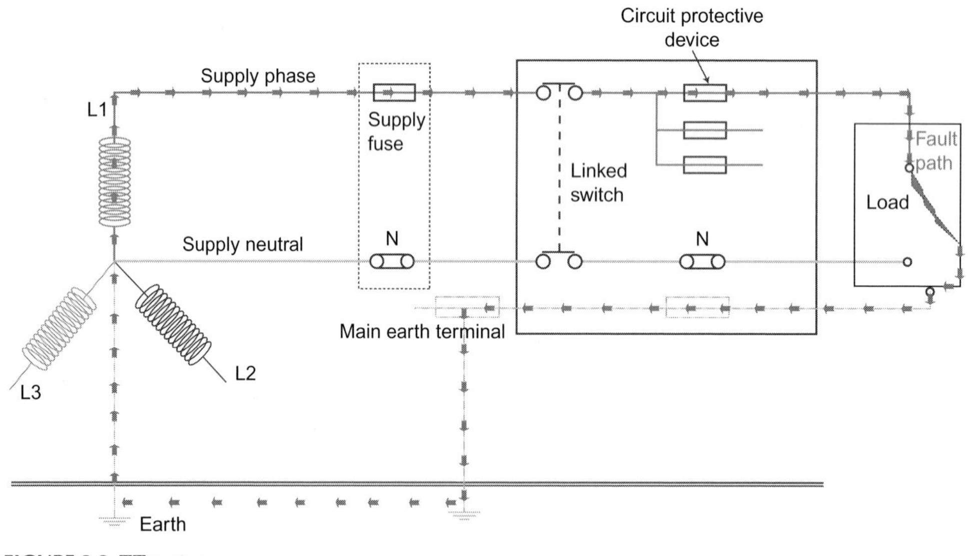

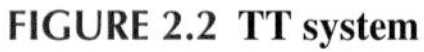

FIGURE 2.2 **TT system**

As can be seen in Figure 2.2, this type of supply system has an earth at the supply transformer and requires an earth to be supplied by the consumer by the use of an electrode; hence, TT has two points of earth. Care must be taken when identifying this type of system as just because it is an overhead supply it does not mean that it is TT. Many of these systems have been converted to TN-C-S systems. A TT system

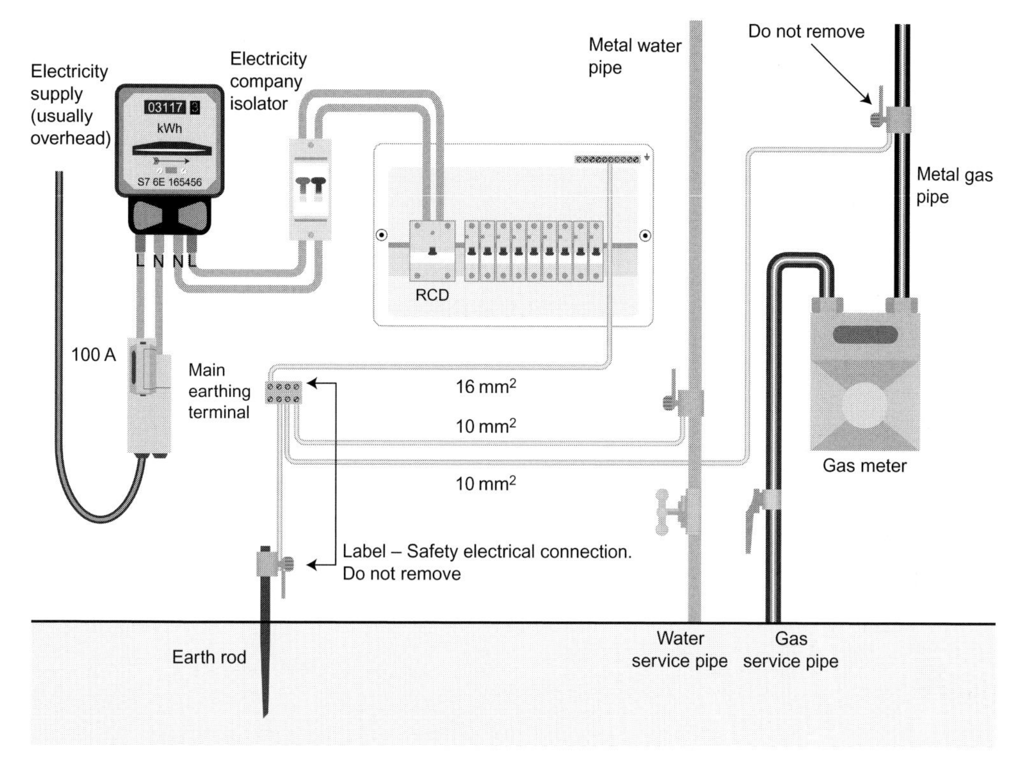

FIGURE 2.3 **TT service head. Reproduced with permission of IEE**

can usually be identified at the service head as there will not be an earthing conductor connected to it or the cable entering it. See Figure 2.3.

TN-S SYSTEM

Probably the most common system that we will come across at this time will be a TN-S system (Figure 2.4). This is generally an underground supply with the supply providing an earth. It can be recognized by the earthing conductor being connected to the metallic sheath (generally lead) of the supply cable (Figure 2.5). The letters TN-S are (T) earth, (N) neutral and (S) separate throughout the whole system.

TN-C-S SYSTEM

Most new systems will be TN-C-S and these will eventually become the most common. In this system, the neutral of the supply is used as the neutral and the earth and the neutral is known as the PEN conductor. Figure 2.6 shows a type of system that is also known as a protective multiple earthing system (PME) (Figure 2.7).

The other systems are TN-C and IT.

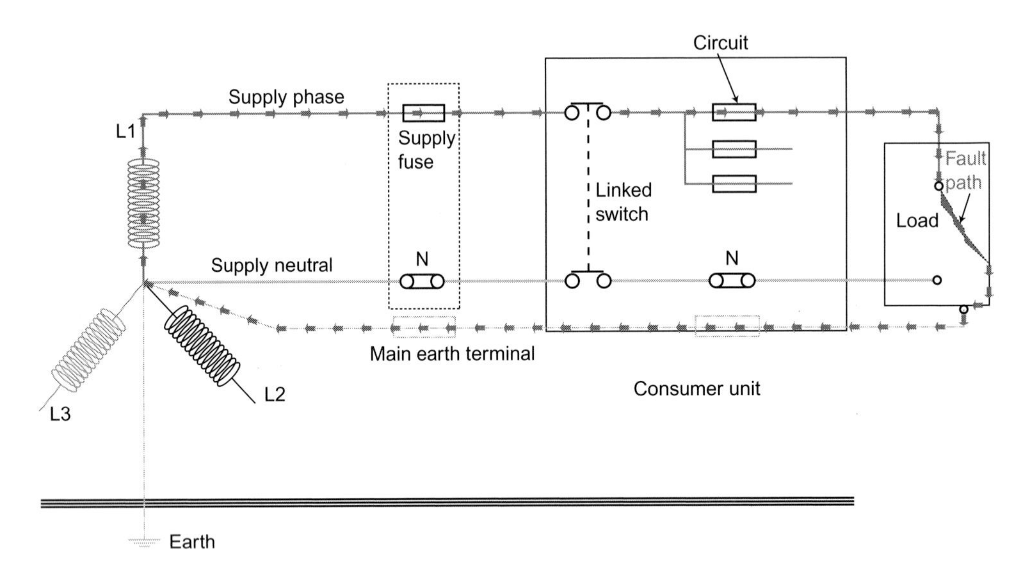

FIGURE 2.4 TN-S system

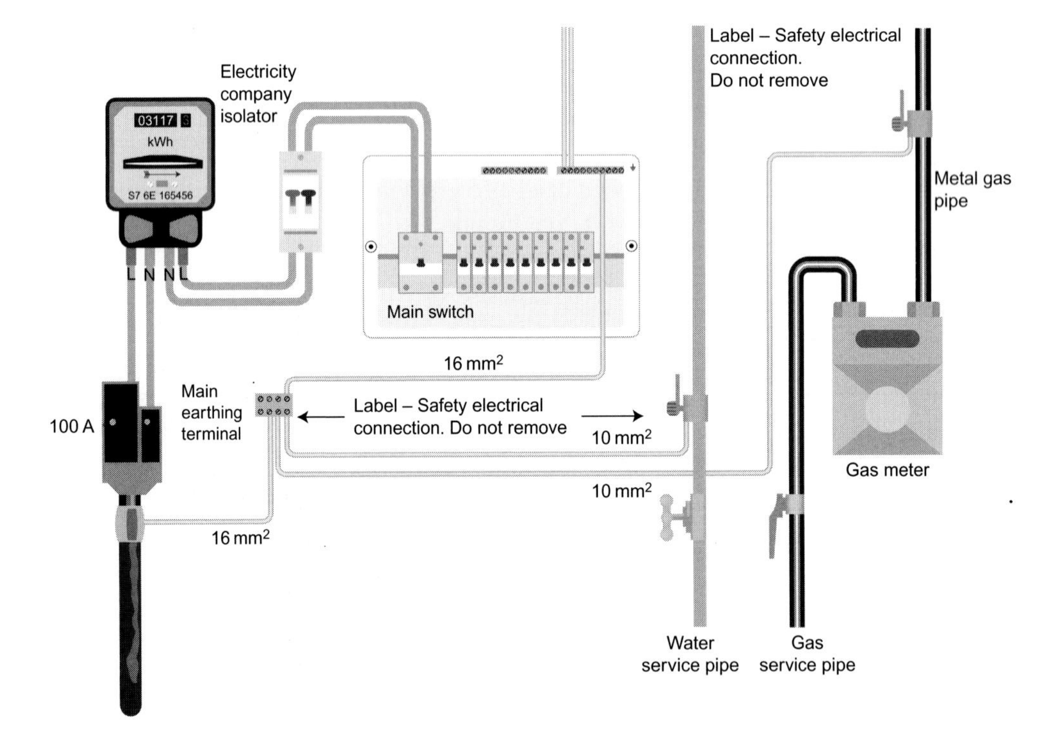

FIGURE 2.5 Service head. Reproduced with permission of IEE

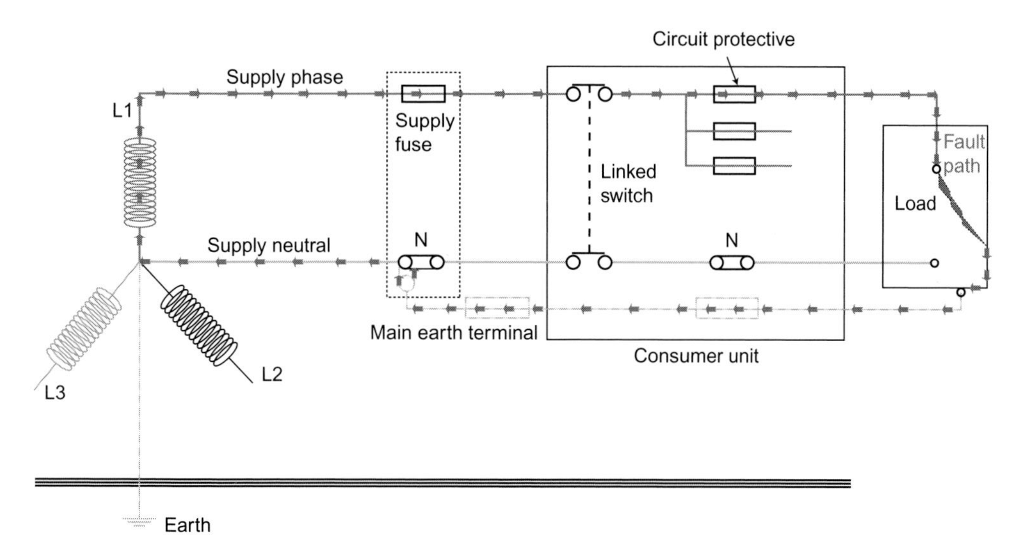

FIGURE 2.6 TN-C-S system

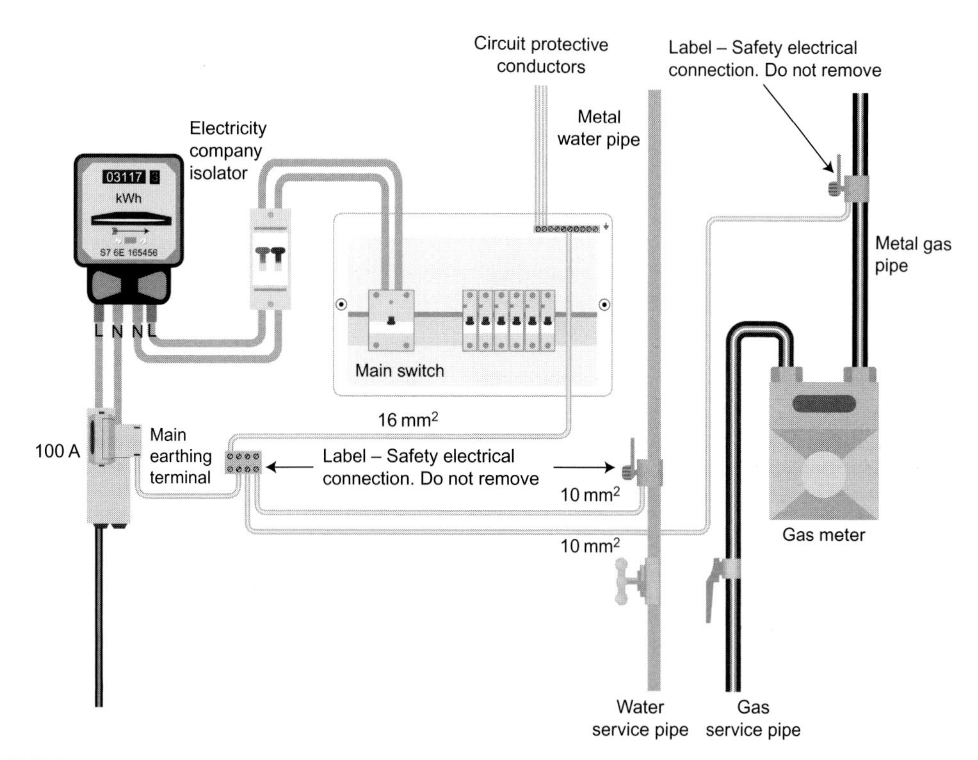

FIGURE 2.7 TN-C-S service head. Reproduced with permission of IEE

A TN-C system uses the neutral throughout the whole of the installation as an earth and is very rarely used as it is prohibited by regulation 543.4.1 and ESQCR 2002 without permission being obtained. An IT system is not to be used in public distribution networks in the UK as it would not comply with the requirements of ESQCR 2002.

Part 2 also contains a list of symbols, which will be explained as required throughout this book.

Chapter 3

Part 3

In this Chapter:

301 ASSESSMENT OF GENERAL CHARACTERISTICS

During the design stage of any installation, careful consideration has to be given to the purpose for which the installation is intended and the type of supply that would be required for the installation.

Chapter 31 provides us with a list of some of the requirements that must be met and some information that must be available before the design can be started.

CHAPTER 31

311 Maximum demand and diversity

The amount of current that is going to be needed in an installation must be known before commencing work. This is to ensure that the supply is large enough.

We need to know if there was a requirement for a three-phase supply due to the type of equipment to be installed, or due to a single-phase supply not having enough capacity for the required demand.

Quite clearly within any installation it would be very unusual for all of the circuits to be used at the same time, and even if they were it would be very unlikely that they would be fully loaded. When we are calculating the size of the required supply we can take this into account by using diversity.

Guidance on diversity is given in guidance note 1 and appendix 1 of the on-site guide. It must be remembered that this is only guidance and to use diversity correctly a good knowledge of the requirements of the installation is needed.

A Practical Guide to the 17th Edition of the Wiring Regulations. DOI: 10.1016/B978-0-08-096560-4.00003-5

Example

A typical house with a 100-A supply may contain the following circuits:

Electric shower 10 kW	50 A
Cooker	32 A
Ring 1	32 A
Ring 2	32 A
Immersion heater	16 A
Upstairs lighting	6 A
Downstairs lighting	6 A

If we were to add up the requirements of each circuit, we have a total demand of 174 A but the main fuse never blows due to overload. This is because the circuits are rarely loaded to their full capacity, and even if they were it would be very unlikely for all of the circuits to be in use at the same time. This is really the principle of diversity.

Using the tables in guidance note 1 and appendix 1 of the on-site guide we can note the following.

The shower, which would be classed as an instantaneous water heater, would have no diversity allowable. However, only the rated load of the appliance should be used, not the rating of the protective device.

$$\text{Shower: } \frac{10{,}000}{230} = 43.47\text{ A}$$

The cooker circuit already would have diversity applied in most cases. For instance, if the cooker had the following:

1 × 4 kW oven
2 × 1.5 kW hob rings
1 × 2 kW hob ring
1 × 1 kW hob ring
1 × 2 kW grill

$$\text{Total power required} = 12\text{ kW}$$

$$\text{Total load (A)} = \frac{12{,}000}{230} = 52\text{ A}$$

It is not likely that all of the elements of the cooker will be used at the same time. You may think that all of the cooker will be used, perhaps on Christmas day and other special occasions. Of course this is true, but usually due to the required cooking times of the food the oven will be on first, and it will have reached its required temperature before the hobs are used. The cooker will have thermostats and simmer stats built into it which will switch on and off to maintain the cooking temperatures.

This will mean that for short periods of time an overload could occur. It would be perfectly acceptable for this circuit and most other circuits for a small overload to occur occasionally. This is another reason why installed equipment must comply with the relevant British Standard.

The guidance notes for BS 7671 suggest that a small overload is between 1.25 and 1.45 A. **Protective devices** have to comply with certain requirements; for example:

BS 88 fuses must be able to carry a current of at least 1.25 times the fuse rating for a duration of 1 hour without fusing. However, they must fuse within 1 hour at 1.6 times their rating.

BS EN 60898 and BS EN 61009-1 protective devices must carry a current of at least 1.13 times the device rating for 1 hour without operating. However, they must operate within 1 hour at a maximum of 1.45 times their rating.

The 1 hour is known as the conventional time and applies to protective devices up to 63 A. For protective devices of 63 A or above, the conventional time is extended to 2 hours.

The current which causes operation of the protective device in the conventional time is shown by the symbol I_2.

Diversity in this installation can now be applied.

Ring circuits: Diversity is calculated at 100% for circuit 1 and 40% for circuit 2.

Ring 1	32 A
Ring 2	32 A × 40% = 12.8 A

$$\text{Total allowance for rings} = 44.8\ \text{A}$$

Immersion heaters are thermostatically controlled water heaters. No diversity is allowable for these, although only the total load should be calculated.

$$\text{Immersion heater: } \frac{3000}{230} = 13\ \text{A}$$

Lighting can be calculated using 66% of the total demand.

$$\text{Two lighting circuits at } 6\ \text{A} = 12\ \text{A}$$

$$12 \times 66\% = 7.92\ \text{A}$$

Now the maximum demand for the installation is:

Shower	43.37 A
Cooker	32 A
Ring circuits	42.8 A
Immersion heater	13 A
Lighting	7.92 A
Total maximum demand	139.09 A

As you can see from the calculation, it still looks as though the maximum demand will be greater than the rating of the supply protective device. This situation is not unusual.

Within most domestic installations the ring circuits are not fully loaded, with any large loads such as washing machines and dishwashers being used intermittently. Showers are used for a very few minutes each day and it is unlikely, unless you have children like mine, that the lights are all on at once. This, along with the cooker as explained earlier, is why diversity can be used. With this in mind, it can be seen that this installation is unlikely to suffer from overload.

Many domestic installations will have a far greater load than the example used here and still be protected by a main fuse of 100 A without encountering any problems.

Although the tables in guidance note 1 and the on-site guide are extremely useful, the values used are by no means cast in stone.

Where diversity is required to be assessed for large residential blocks, commercial, industrial and other installations, it can be assessed by a competent person using knowledge, experience and common sense.

312 Arrangement of live conductors and type of earthing

The design of any installation will require the designer to have information regarding the type of supply – is it to be three or single phase? When the supply is provided by a distributor, we need to know the type of supply system which is to be or has been provided – is it TT, TN-S or TN-C-S? This information could influence the design of some parts of the installation. The section of this book dealing with chapter 54 provides further information on this.

313 Supplies

We need to know the following.

Voltage

This could be measured or obtained by enquiry. Usually we will know as most of our installations are 230 or 400 V.

Current and frequency

Is the current a.c. or d.c.? If we are working on an unusual installation, we need to obtain this information by enquiry. We know by experience that our normal supply is a.c. with a frequency of 50 Hz.

Prospective short-circuit current (PSCC) at the origin of the installation

This is the maximum current which could flow between live parts as close as we can measure to the meter. It must not be confused with earth fault current.

For single-phase supplies where the protective device for the supply is no greater than 100 A, a fault current of 16 kA may be assumed provided that the current rating of individual protective devices does not exceed 50 A. The consumer unit complies with BS 5486.13 or BS EN 60439-3 and the supply protective device is a BS 1361 type 2.

For three-phase 400-V supplies engineering recommendation P 26 should be consulted, as the magnitude of the fault current will depend on the size of the supply conductors and the length of the supply cable.

The Electricity Safety Quality and Continuity Regulations 2002 (ESQCR) require distributors to provide the value of the PSCC at the supply terminals. A call to the service provider will be one option; the quoted value given will probably be 25 kA as this is as high as it is likely to be.

On a supply that has already been installed the value can be measured; however, the designer should be aware that this value could change if the supply was upgraded locally due to increased demand. Where the measured fault current is exceptionally low, it is a good idea to install protective devices with a higher prospective fault current (PFC) rating than is required. This will ensure continued protection should this value increase.

External earth fault loop impedance, Z_e

This is the resistance between live conductors and earth. It can be obtained by measurement, enquiry or calculation. Where this value cannot be measured, for instance, at the design stage of a new building where the supply has not yet been installed, enquiry is the only option. For supplies which are 230/400 V, which are protected at the supply cut-out by fuses to BS 1361 type 2 or BS 88-2 or BS 88-6, the value of Z_e quoted will usually be:

TT	21 Ω
TN-S	0.8 Ω
TN-C-S	0.35 Ω

This information can be found in the on-site guide.

While calculations for cable sizing, etc., can be carried out using these values, it is important to measure these values before allowing the installation to be put into service. The only system for which the calculation of Z_e would be possible is a TN-C-S system. This is because the fault path external to the installation is the same for both PSCC and prospective earth fault current (PEFC). To enable this calculation to be carried out, the value of PSCC between line and neutral should be known. The calculation is then a simple Ohm's law exercise:

$$\frac{U_0}{Z_e} = \text{PSCC}$$

Maximum demand

Is the supply going to be large enough for the required load? It is always a good idea to allow for future additions to the installation where possible.

Type and rating of the supply protective device

Is the supply fuse large enough for the requirements of the installation? Will it be capable of carrying the PFC without physical damage? (Apart from the melting of the fuse.)

All of the information required for supplies is also required for any source which is to be used for a safety service and standby system. (Regulation 313.2.)

314 Division of installation

In the event of a fault, it would be very dangerous and very inconvenient if the whole of the installation were cut off from the supply. By dividing the installation into circuits the likelihood of this happening is considerably reduced. Separate circuits also provide us with a means of isolation for a section of an installation; this is important when we are required to carry out any maintenance, repairs or inspection to a system as we can work on one circuit at a time and cause a minimum of inconvenience to any occupants.

Further consideration must be given to the use of residual current devices (RCDs). If one RCD were to be used for the protection of the whole system, compliance with the regulations could not be met (regulation 314.2). To gain compliance with this regulation, a minimum of two RCDs would be required and even then consideration should be given as to how the installation is split. For instance, if a dwelling were split with the circuits upstairs on one RCD and the circuits downstairs on another, a fault on one circuit would result in the loss of energy on one complete floor. This could result in a dangerous situation, particularly at night. It would be far better to have the upstairs lighting protected by the same RCD as the downstairs power and the downstairs lighting and upstairs power protected by another RCD.

Of course, by far the best option is to provide each circuit with its own dedicated RCD protection by using residual current breakers with overload protection (RCBOs). This option is probably the most expensive but would certainly provide compliance with section 314 of BS 7671.

CHAPTER 32

Classification of external influences

An external influence is anything that could affect the electrical installation which is not part of the installation. This is not worthy of being called a chapter as it simply directs you to chapter 51 and appendix 5.

The three categories given in appendix 5 are the following.

Environment

Where is the installation going to be? Will it be affected by water, foreign bodies, temperature and humidity, altitude, corrosion, impact, vibration, mechanical stresses, flora (plants), fauna (animals), electromagnetic, solar or seismic (earthquake) activity, lightning, or movement of air and wind?

All of these environmental conditions must be protected against where it is likely that they could have an adverse affect on the installation.

Utilisation

What is the installation going to be used for?

What are the **capabilities** of the persons using the installation? Are they children? Are they handicapped? What is the likelihood of the persons using the installation being in **contact with earth?** How easy would **evacuation** be and what risk do the **materials** present? Are they explosive or poisonous? Is there a fire risk?

Construction

What is the building made of – is it **combustible or non-combustible**? Will it **propagate the spread of fire** due to the building shape? Will the building **move**? – possibly due to its size or being connected to another building of a different construction. Is the building **flexible or unstable?** – it could be a tent or a marquee or even a false ceiling.

Section 512.2 of BS 7671 requires that any installed equipment is suitable for any external influence by which it may be affected.

CHAPTER 33

331.1 Compatibility

Consideration must be given to any equipment which could affect other parts of the installation. This could be due to voltage drop caused by large motors starting or large welding plants which may cause the load to fluctuate rapidly. Unbalanced loads could have an adverse effect on the installation.

332 Electromagnetic capability

In most cases, provided we install equipment which has been manufactured to British Standards, this regulation will be catered for, although in some cases surge protection may be required. Consideration must also be given to electromagnetic interference of other parts of the installation.

CHAPTER 34

341 Maintainability

All electrical installations should be as simple as possible and all components that may need to be inspected or worked on must be accessible. Where the means of access is not immediately obvious, operating instructions should be positioned near to the component where they can be seen.

Adequate working space must be provided where working on or near electrical equipment may give rise to danger. Consideration must also be given to the positioning of luminaires so that they can be safely re-lamped or repaired by maintenance personnel.

CHAPTER 35

351 Safety services

A safety service could be emergency escape lighting, fire-alarm systems, fire pumps or any system that is put in place to protect or warn persons in the event of a hazard.

These types of system are often required by statutory authorities and may need to be in place before a particular type of licence is granted, for example escape lighting in a club or theatre.

This type of service has to have a standby supply. The source of the supply can be storage batteries, primary cells, generator sets which are separate from the supply, or even a completely separate supply from the supply network. The section of this book dealing with chapter 56 provides more information on this.

CHAPTER 36

361 Continuity of service

For any circuit which would present a risk if there were to be a power failure, such as a life-support system, consideration must be given to:

- The type of system earthing.
- Discrimination of protective devices to ensure that where possible one fault does not disconnect more than one circuit. It could be very dangerous if one circuit fault operated the protective device protecting the whole distribution board.
- The number of circuits required.
- Will multiple power supplies be required?
- Should monitoring devices be used?

Part 4

In this Chapter:

Part 4 provides the requirements for protection for safety and consists of four chapters.

Chapter 41 provides information on the requirements for protection against electric shock. The fundamental rules of protection against electric shock are that hazardous live parts should not be accessible, and accessible conductive parts should not be dangerously live when in use without a fault or when a single fault occurs within an installation. In other words, all live parts must be suitably enclosed, and when a fault occurs any person or livestock must not be at risk of electric shock from exposed metalwork.

Enclosures and insulation are classed as **basic protection** as they prevent unintentional contact with live parts.

Fault protection prevents exposed or extraneous conductive parts becoming hazardous under fault conditions. In any installation, we are required to provide both basic and fault protection.

CHAPTER 41

Protection against electric shock

Chapter 41 provides a safety standard which protects persons and livestock.

Clearly, it is important that in any installation, persons and livestock are prevented from coming into contact with any live parts. This does not mean that we have to fit accessories and equipment with tamper-proof screws and fixings, although in some environments such as prisons this may be desirable.

This type of protection is known as BASIC protection and consists of insulation, barriers and/or enclosures. Obstacles and placing out of reach can also be used

A Practical Guide to the 17th Edition of the Wiring Regulations. DOI: 10.1016/B978-0-08-096560-4.00004-7

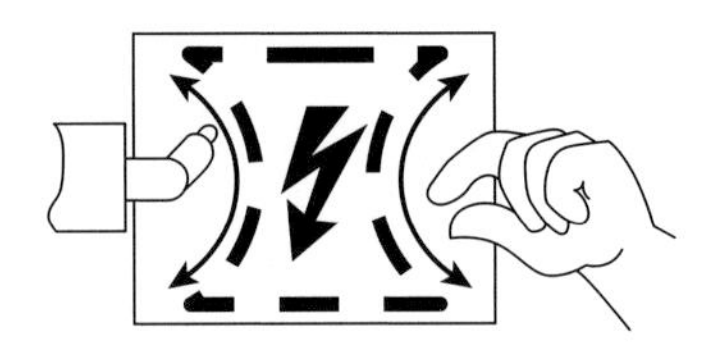

FIGURE 4.1 **IPXXB. Test finger penetration to a maximum of 80.0 mm must not contact hazardous parts**

in some instances, although these would not be used unless there were special circumstances. Placing out of reach would only be used to prevent unintentional contact with live parts and would normally be found where overhead lines are used between buildings. When placing out of reach is used, it must be installed to the standard set out in the ESQCR 2002.

Where barriers or enclosures are used to prevent contact with live parts, they must comply with regulation 416.2. Live parts must be inside enclosures. The sides, front, bottom and internal barriers can have a minimum degree of protection of IP2X or IPXXB.

For compliance with IP2X, a hole in an enclosure must not allow a sphere of 12.5 mm diameter to pass through it. IPXXB is known as finger protection. For compliance with this, where there is an opening in an enclosure it must not be possible for a person to touch a live part through the opening (Figure 4.1).

Clearly, sometimes this is impossible to achieve, particularly where the replacement of parts is required. Such items include lampholders with the lamp removed and some fuses (regulation 416.2.1). As far as the Wiring Regulations are concerned, this is not a problem as long as suitable precautions are in place to prevent unintentional contact. Persons are made aware that live parts can be touched through the opening and that they should not be touched intentionally. This could be achieved by the use of a printed label or sign. Care should also be taken to ensure that the opening is as small as possible but large enough to allow safe replacement of a part and proper functioning of equipment.

The top surface of barriers and enclosures must have protection of no less than IP4X. This is a hole with a diameter of no greater than 1 mm or IPXXD, which is protection against wire or probes. This requirement only applies to the top surfaces which are readily accessible and would not apply to enclosures which are fitted reasonably close to a ceiling or even a shelf. It applies to the top surface of any surface accessory and switch or socket outlet boxes.

In some instances, it may be necessary to open an enclosure or remove a barrier for replacement or adjustment of parts. In this case precautions must be taken to prevent easy access to unskilled persons (regulation 416.2.4). This can be achieved by providing a door or barrier which can only be opened using a key or tool, or providing an interlock to ensure that the enclosure cannot be opened until the supply is isolated. An example of this would be an isolator (*you will be aware that you cannot open the door of an isolator when it is switched on*).

All distribution boards and consumer units that comply with BS EN 60439 will satisfy this regulation as they have intermediate barriers to IP2X or IPXXB which are only removable using a key or tool. It is important to remember that if for any reason a capacitor is installed in an enclosure behind a barrier, then a notice must be fitted to warn any person removing the barrier that a dangerous charge may be present after the equipment has been switched off (regulation 416.2.5). This does not apply to small capacitors used for relays and the like.

Clearly, all electrical installations must have basic protection; this alone would be unsuitable as protection against electric shock as it is only used to provide protection where there is no fault. It is very important that in any installation the best protection possible is provided by good design and installation. This includes the provision of measures designed to protect persons, property and livestock when a fault occurs. A combination of basic and fault protection is the most common method of achieving compliance with the Wiring Regulations.

Fault protection would be achieved when a circuit or even a complete installation was automatically turned off when a fault occurred. This method of protection is called ADS.

ADS requires several important parts of an installation to be in place for it to be relied upon for satisfactory protection.

Each circuit must have a suitable protective device and a means of earthing must be present – for a TN system this is the supplier's earthing arrangement and for a TT system this is an earth electrode along with the installation of a suitable RCD.

Each circuit must have a suitable circuit protective conductor (CPC) and all extraneous conductive parts must be bonded. Supplementary bonding may also be required.

Let us look at each of these requirements individually.

Protective devices

Protective devices must be selected to suit the requirements of the circuit and the load which they are intended to protect. Is it going to be a fuse or a circuit breaker? To comply with regulation 411.3.2.2, any final circuit of 32 A or less must disconnect under fault conditions in a maximum of 0.4 s if the supply is a TN system or 0.2 s if it is a TT system. Distribution circuits (sub-mains) and circuits rated above 32 A can have their disconnection time extended to 5 s.

Note: Circuits feeding fixed equipment used in highway power supplies can have a 5-s disconnection time regardless of the circuit current rating.

If circuit breakers are going to be used for protection, disconnection times will automatically conform to the requirements provided that the circuits meet the Z_s values given in table 41.3. The product standard for circuit breakers to BS EN 60898 requires that they operate within 0.1 s. For a type B circuit breaker, the time must be achieved when a current of between 3 and 5 times its rating passes through it. Type C requires a current of between 5 and 10 times its rating and a type D requires a current of between 10 and 20 times its rating. The symbol for the current causing automatic disconnection of the device is I_a.

To be able to produce tables for Z_s values, a specific value of current is required and the worst case is used. This is 5 times the rating for a type B, 10 times for a type C and 20 times for a type D.

When these values are used it is simply a case of applying Ohm's law to the supply voltage (230 V) and I_a to obtain a Z_s value for the device. This calculation will satisfy regulation 411.4.5 ($Z_s \times I_a \leq U_0$).

Example

Type B

The Z_s for a 20-A type B can be calculated.

$$5 \times 20 = 100$$

$$\frac{230}{100} = 2.3\Omega$$

Type C

$$10 \times 20 = 200$$

$$\frac{230}{200} = 1.15\Omega$$

Type D

$$20 \times 20 = 400$$

$$\frac{230}{400} = 0.57\Omega$$

This calculation is fine for BS EN circuit breakers; however, there is a considerable number of BS 3871 circuit breakers still in use. The calculation used earlier is still used to calculate Z_s.

These types of circuit breaker are identified by the use of numbers as shown in Table 4.1.

TABLE 4.1 Circuit breakers

Type	Multiplier
1	4
2	7
3	10
4	20

Example

The Z_s value for a 20-A type 1 BS 3871.

$$4 \times 20 = 80$$

$$\frac{230}{80} = 2.87\,\Omega$$

Remember these protective devices will all operate within 0.1 s; therefore, disconnection times of 0.4 and 5 s do not need to be considered.

The same calculation can be used for all types of fuse. To obtain the correct operating current for use in the calculation the tables in appendix 3 must be consulted. For these types of protective device different disconnection times must be considered depending on what they are being used for. The section dealing with chapter 41 in this book provides further information on this.

When we look at disconnection using an RCD, regulation 411.5.3 provides us with the calculation $R_A \times I_{\Delta n} \leq 50\text{V}$.

R_A is the total resistance of the earth electrode and the protective conductor.
$I_{\Delta n}$ is the rated operating current of the RCD.

This calculation is used where an RCD is used in a TT system, but remember that where an RCD is used for protection on a TN system we need to use the formula$Z_s \times I_a \leq U_0$.

Z_s is the earth loop impedance.
I_a is the current causing automatic disconnection.
U_0 is the nominal voltage to earth.

In the case of an RCD being used on a TN system, I_a is the current causing automatic disconnection, which is the trip rating of the RCD ($I_{\Delta n}$). If we were using a 30-mA RCD the maximum Z_s would be:

$$\frac{U_0}{I_a} = Z_s \quad \text{or} \quad \frac{230}{0.03} = 7666\,\Omega$$

It is always better not to use this calculation but to install the circuit to comply with the tables for Z_s wherever we possibly can.

Protective earthing

Every low-voltage circuit that is installed must have a CPC which must be terminated at each point in the wiring system and at each accessory.

Each exposed conductive part must be connected to the earthing system, either individually or all together. Where a twin and earth wiring system is used, the CPC of each circuit provides an individual earth fault path, whereas if a steel conduit or trunking system is to be used then the conduit and trunking provide a common earth path for a number of circuits. In all cases, the protective earthing must comply with chapter 54, which is covered in Part 5 of this book.

TABLE 4.2 Maximum permissible lengths of copper conductors

Size (mm^2)	Length (m)
10	27
16	43
25	68
35	95

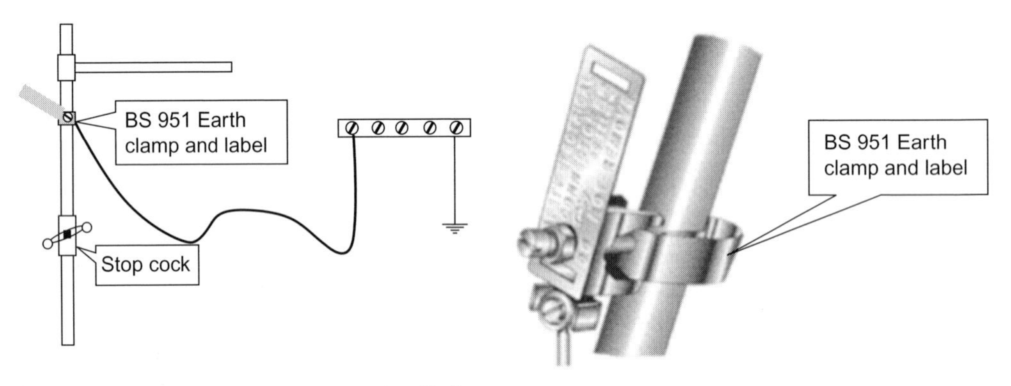

FIGURE 4.2 **Protective bonding to water installation**

Protective equipotential bonding

All extraneous conductive parts within an installation must be connected to the main earthing terminal. Main protective bonding conductors are used for this. They must be selected to comply with chapter 54, and when measured from end to end, the resistance of the conductor must not exceed 0.05 Ω. Table 4.2 gives maximum permissible lengths for copper conductors to ensure the maximum value is not exceeded.

Extraneous conductive parts include the following.

(i) Water pipes as close to the point of entry as possible on the consumer's side of the stop cock (Figure 4.2).

(ii) Gas pipes as close to the point of entry as possible but on the consumer's side of the meter. In the case of a meter sited outside, the gas can be bonded within the outside enclosure. The bonding conductor must enter the enclosure using a separate hole, not be pushed through with the gas pipe (Figure 4.3).

(iii) Any other installation pipework or ducting that may come into contact with exposed conductive parts. This includes central heating systems and air conditioning.

(iv) Exposed metallic structural parts of a building. These do not include steel lintels and the like, but do include any steelwork that is in contact with exposed

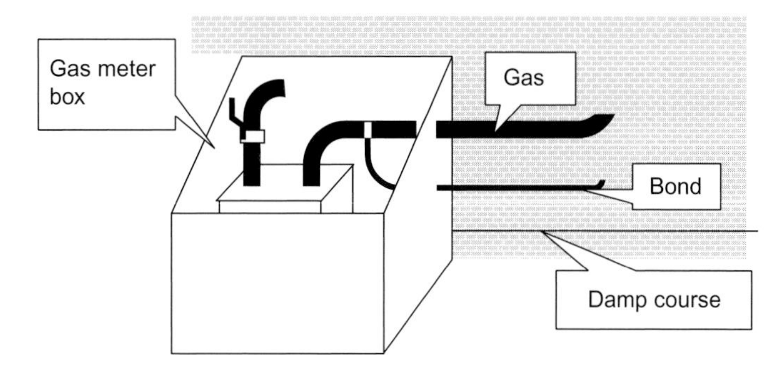

FIGURE 4.3 **Protective bonding to gas installation**

conductive parts or the general mass of earth. This would not apply to metal suspended ceiling systems; in most cases where the ceilings incorporate class I light fittings the ceiling will be unintentionally in contact with earth through the light fittings and would be unlikely to introduce a potential. Bonding of a suspended ceiling would be pointless in most cases as the electrical integrity of the ceiling joints could not be relied upon, particularly over large areas.

(v) Lightning protective systems should be considered with regard to bonding. Unfortunately, it is not just a matter of connecting the correct size of bonding conductor to the lightning conductor, unless the person carrying out the work has a full understanding of BS EN 62305. The advice of a consultant is required. A letter must be obtained stating that the lightning conductor must be bonded, or in the case of no bonding being required the letter must state that this is the case.

Remote buildings on the same installation need to be bonded individually and must comply with the same requirements as the main building. The section on chapter 54 (Part 5) explains this in more detail.

Automatic disconnection of supply

A requirement of the regulations is that in the event of a fault occurring of negligible impedance between live conductors, or line conductors and earth, the supply is interrupted within the times stated in table 41.1 of the Wiring Regulations. For a TN supply with a U_0 of 230 V, this is 0.4 s for any final circuit with a current rating $\leq$32 A and 5 s for any final circuit >32 A.

A TT system will require a disconnection time of 0.2 s for any circuit rated $\leq$32 A and 1 s for any circuit >32 A.

(U_0 is the voltage between line and earth; therefore, U_0 would be 230 V for a 400-V three-phase system.)

As stated earlier, distribution circuits can have their disconnection times extended to 5 s for a TN system or 1 s for a TT system. Of course, this is only possible where fuses are used.

It is possible that some confusion could arise over the disconnection times for circuits on TT systems. In all but very extreme cases the only way that the 0.2-s disconnection time could be met is by the use of an RCD. Where general-purpose RCDs to BS 4293 are used, this is not a problem as they must disconnect within 200 ms (0.2 s). Clearly, this would satisfy regulation 411.3.2.2.

The requirement for BS EN type RCDs is that they disconnect within 300 ms (0.3 s). This disconnection time is not fast enough to satisfy the regulations, and their use could cause some concern. However, the regulations do take into account that a fault current is going to be at least twice the trip rating ($2_{\Delta n}$) of the RCD. As a fault current is almost certainly going to be $2 \times I_{\Delta n}$, a BS EN type RCD will comply with the requirements.

The product standard for BS EN type RCDs is given on the first page of appendix 3, in table 3A of the Wiring Regulations. It shows us that if it operates within the correct time (300 ms) at times one $I_{\Delta n}$, then it will operate at 150 ms or less at twice $I_{\Delta n}$.

Therefore, if the RCD is satisfactory when tested at times half, times one and times five (5 times for 30 mA and below only) no further tests are required, and it can be accepted that the RCD is satisfactory.

Table 41.5 of BS 7671 provides maximum values of Z_s which will ensure the correct operation of RCDs; however, the resistance of the actual earth electrode (R_A) should not be more than 200 Ω. This is because any resistance value above 200 Ω would be unreliable. It is recommended that the resistance of earth electrodes is measured under the worst conditions; of course this does not mean the worst conditions for us (raining, cold and windy). Earth electrode resistance must be measured under the worst conditions for the electrode, which would be as dry as possible. This would ensure that the highest resistance value was measured and recorded. If the measurement was taken when soil was wet, it may dry out and the electrode resistance could rise considerably and prevent it from operating correctly.

The maximum values shown in table 41.5 are obtained by the use of Ohm's law. The calculation is given in regulation 411.5.3 as:

$$R_A \times I_{\Delta n} \leq 50\,\text{V}$$

R_A is the resistance of the earth electrode and the protective conductor (Z_s can be used in place of R_A).

$I_{\Delta n}$ is the trip current rating of the RCD.

50 V is the maximum permitted touch voltage.

When carrying out this calculation, it is important to remember that in most cases the trip rating of an RCD is given in milliamps; therefore, the calculation to find the maximum resistance permissible for a 30-mA RCD is as follows: 50/0.03 = 1666.66 rounded up to 1667 Ω.

(*This calculation is for theory purposes only and has no practical use.*)

Additional protection

To comply with regulation 411.3.3, socket outlets not exceeding 20 A must be provided with RCD protection. This also applies to mobile equipment up to 32 A which is used outdoors. The regulations tell us that exceptions are permitted for socket outlets if they are used under the supervision of skilled or instructed persons, or if they are specifically labelled or identified as being provided for a particular item of equipment. Unfortunately, it is not quite as simple as it seems; this is because to gain compliance with BS 7671 the regulations must be read in their entirety.

As we will see when we get to chapter 52 of BS 7671, cables which are installed in walls but which are not encased in earthed metal or 50 mm deep will also need to be RCD protected. This means that almost all circuits installed in domestic environments will require RCD protection, as will many circuits in commercial and industrial situations regardless of whether they supply socket outlets or not.

Many customers will not be too keen on having their freezers protected by RCDs as they will be concerned about nuisance tripping. This problem can be overcome by installing the cable in a length of screwed conduit buried in the wall with a bush and coupler connecting it to the metal knockout box. The box and conduit can then be earthed by the use of a fly lead to the box from the socket. The socket will need to be on a radial as all of the other sockets will need to be RCD protected (Figure 4.4).

In some instances to achieve compliance with regulation 411.3.3, it will be possible simply to fit an RCD protected socket outlet to BS 7288, particularly where the installation is in surface conduit or trunking; however, it will not comply with regulation 522.6.5/6 (concealed cables).

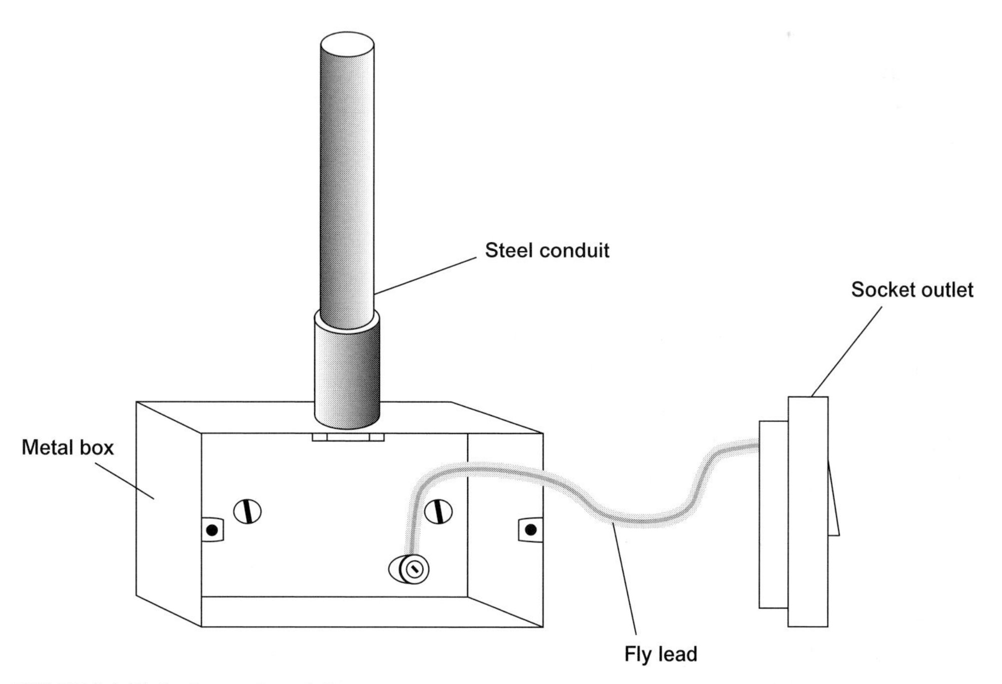

FIGURE 4.4 Fly lead to steel conduit system

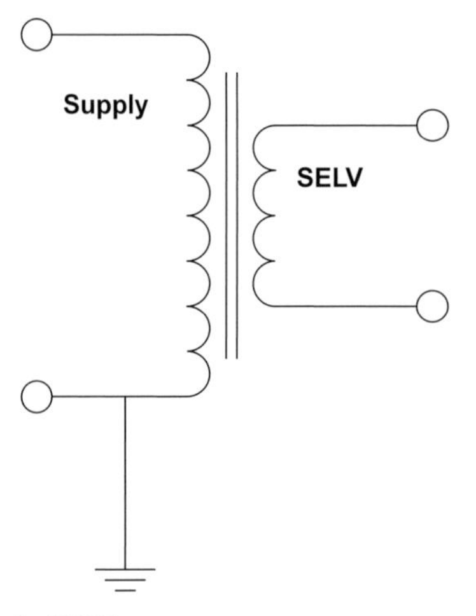

FIGURE 4.5 **Isolating transformer for SELV**

411.7 Functional extra low voltage

Functional extra low voltage (FELV) would be a voltage of up to 50 V a.c. or 120 V d.c. and would not be used as a means of protection on its own. Unlike SELV or PELV, it must always fulfil the requirements for basic and fault protection as if it were a normal low-voltage circuit.

FELV is a system normally supplied by a transformer which would not meet the requirements for SELV or PELV due to there being insufficient insulation between the FELV supply and circuits supplied at a higher voltage. If FELV is used for any reason, it is a requirement that any socket outlet, plug or coupler is not compatible with any other device used within the same premises (regulation 411.7.5).

Extra low voltage provided by SELV or PELV

Unlike FELV, it is acceptable to use SELV or PELV as a protective measure in all situations with the exception that in certain locations as listed in part 7 of BS 7671 the maximum permissible voltage may be limited to a lower value. An example of this would be current using equipment located in a bathroom. The voltage in this instance is limited to 12 V a.c. or 30 V d.c.

A typical example of the use of SELV is extra-low-voltage lighting using separate transformers for each luminaire or one transformer supplying several luminaries. Alarm systems are also often supplied by SELV.

The difference between a SELV and a PELV system is that a SELV system is supplied by an isolating transformer. The primary side of the transformer is supplied by an earthed low-voltage circuit and the secondary side supplies a load at between 0 and 50 V and is totally isolated from any earthing provided within the installation (Figure 4.5).

A PELV system must meet all of the requirements for a SELV system, but it will be connected to the earthing of the low-voltage part of the installation.

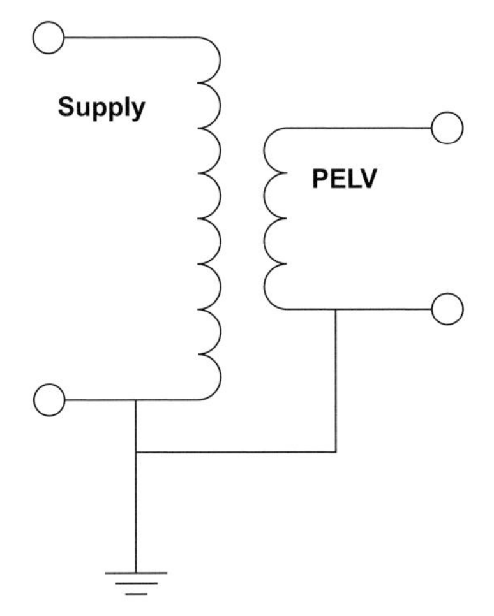

FIGURE 4.6 **Transformer for PELV**

An example of PELV is a Class I supply for a computer (Figure 4.6).

Regulation 414.3 sets out the requirements for SELV and PELV supply sources. The most common source is a safety isolating transformer; other sources are certain motor generators or batteries. In some instances, higher voltages can be used provided that in the event of a fault between live parts and exposed conductive parts, the voltage at the output terminals of the device is immediately reduced to 50 V a.c. or 120 V d.c. An insulation resistance test instrument would meet these requirements.

Where SELV and PELV circuits are used, basic protection is still required for any voltage above 25 V a.c. or 60 V d.c. Basic protection is also required when equipment using these voltages is immersed.

In circuits where the voltage is less than 25 V a.c. or 60 V d.c. which are being used in normal dry conditions basic protection would not be required, and in wet conditions basic protection would be required for these circuits if they were above 12 V a.c. or 30 V d.c. Below this, basic protection is not required (Table 4.3).

411.8 Reduced low-voltage systems

This type of system would normally be a 110-V supply, which if it was three phase would not exceed 63.5 V between line conductor and earthed neutral or 55 V between the earthed mid-point of the transformer to earth. This ensures that only an electric shock of 55 V could be received from any live conductor to earth (Figure 4.7).

Basic protection must be provided for this type of system. Fault protection must also be provided by using overcurrent protective devices, with a disconnection time not exceeding 5 s. RCD protection may also be used.

Maximum Z_S values can be found in table 41.6 of BS 7671. This table can be a little confusing as it states a 5-s disconnection time with a voltage of 55 V. This is a 110-V supply with the mid-point of the supply transformer tapped to earth (standard site transformer).

TABLE 4.3 SELV, PELV and FELV in relation to safe separation and the relation to earth

Type of separation		Relation to earth of protective conductor		Method
Current source	*Circuits*	*Circuits*	*Exposed conductive part*	
Current source with safe separation, for example a safety transformer or an equivalent current source	Circuits with safe separation	Unearthed circuits	Parts may not intentionally be connected to earth or to a protective conductor	SELV
		Earthed and unearthed circuits permitted	Parts may be connected to earth or to a protective conductor	PELV
Current source without safe separation, i.e. a current source with only basic separation, for example a safety transformer to IEC 60989	Circuits without safe operation	Earthed circuits permitted	Parts shall be connected to the protective conductor of the primary circuit of the supply system	FELV

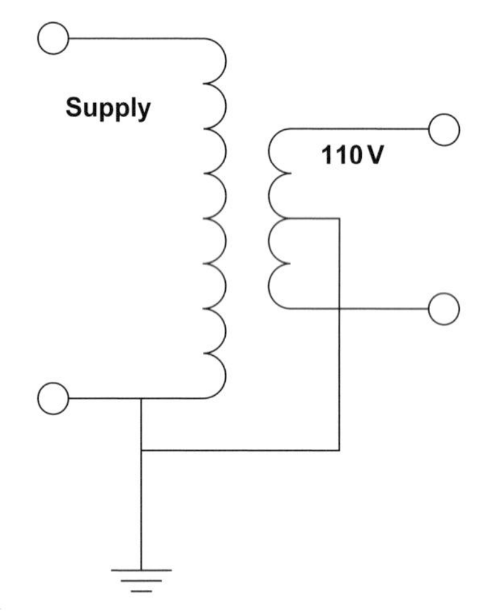

FIGURE 4.7 Reduced voltage transformer (centre tapped)

The Z_s values for circuit breakers given in table 41.6 are calculated in the same way as the circuit breakers in table 41.3. The only difference is that a voltage of 55 or 63.5 V is used instead of 230 V.

Example for a 32-A type B circuit breaker.

$$32 \times 5 = 160$$

$$\frac{55}{160} = 0.34\,\Omega$$

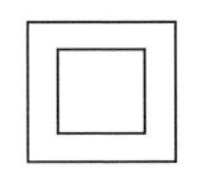

FIGURE 4.8 Class II symbol

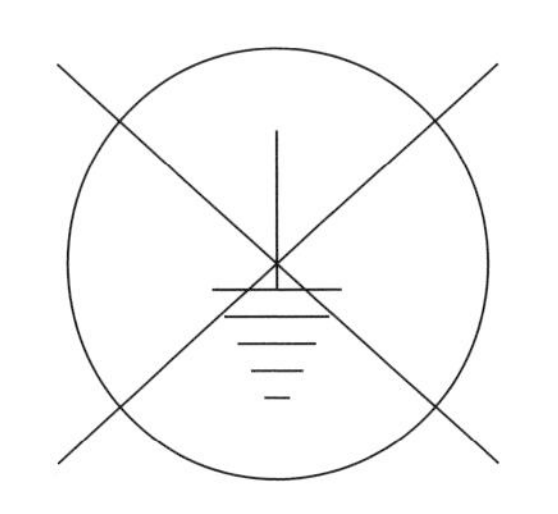

FIGURE 4.9 Do not connect to earth symbol

412 Double or reinforced insulation

Double insulation consists of a layer of basic insulation and supplementary insulation which together provide two layers of insulation between live and accessible parts. Reinforced insulation is one solid mass of insulation or several layers of insulation which provide protection equivalent to double insulation. This type of insulation is found on Class II equipment. Look for the symbol indicating that the equipment is Class II or for the symbol indicating the equipment must not be connected to earth (Figures 4.8 and 4.9).

413 Electrical separation

Electrical separation is covered by two sets of regulations. Regulation 413 is for use where one single item of equipment is supplied from an unearthed source, which is usually an isolating transformer. The principle of this is that the primary side of the transformer is supplied by an earthed low-voltage circuit, with the secondary side being low voltage up to 500 V with no connection to earth. Basic protection is provided by basic insulation, barriers and enclosures in compliance with regulation 416. Fault protection is provided by the use of regulations 413.3.2 to 413.3.6. Great care must be taken to ensure that any exposed conductive parts of the separated circuit and the equipment connected to it do not come into contact with any other earthed parts of other circuits within the installation.

A shaver socket is a good example of electrical separation for a single piece of equipment (Figure 4.10).

Where electrical separation is being used to supply more than one item of equipment, the set of regulations 418.3 must be used.

A notice to comply with regulation 514.13.2 must be fixed in a prominent position at every point of access, warning that the equipment must not be connected to earth. All exposed conductive parts of the separated circuits must be bonded together using insulated bonding conductors. The earthing terminal of socket outlets, although

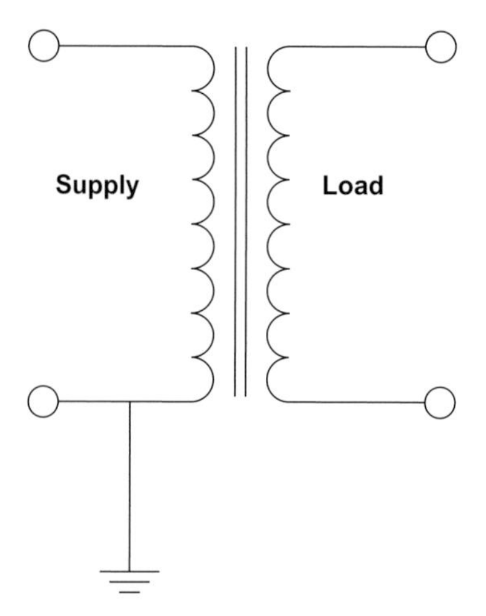

FIGURE 4.10 **Low-voltage isolating transformer**

not connected to earth, must be connected to the equipotential bonding system. All cables must contain a CPC, except when they are supplying Class II equipment. This conductor is to be used as a bonding conductor. Overcurrent protection must be provided on each pole of the separated circuit in case of a two-fault situation.

It is also recommended that the length of the circuit should not exceed 500 m and that the system should not exceed 100,000 volt meters (Vm).

For example: A 230 V circuit must not be longer than

$$\frac{100,000}{230} = 434.78\,\text{m}$$

This type of installation is suitable for electrical repair workshops where it is necessary to work on live equipment for fault finding, repair and adjustment.

417 Obstacles and placing out of reach

The use of these measures of protection will provide basic protection only, and these measures are only for use in installations which are controlled or supervised by skilled persons. They will not be found within most installations.

Figure 417 in BS 7671 gives distances for arm's reach.

418

The following forms of basic protection are only to be used where the installation is controlled or under the supervision of skilled or instructed persons.

418.1 Non-conducting locations

This type of protection is not generally used in Great Britain but was traditionally used overseas where all trades had an understanding of this method of protection.

Individuals in this country who undertake electrical work are governed by the Electricity at Work Regulations (EAWR). We also have to comply with the building regulations. It is an unfortunate fact that persons who want to carry out most forms of construction work do not require qualifications. For this reason, the required support between trades to ensure that this type of protection is effective is unlikely to be available.

418.2 Protection by earth-free local equipotential bonding

These are areas in which most electricians will never be called upon to work, as they are very specialist areas such as operating theatres and laboratories. In some instances, equipment used in these areas may, due to their construction, have exposed conductive parts which if connected to earth will not function correctly. The principle of this type of protection is that the room containing the electrical equipment is completely insulated from earth. All accessible exposed conductive parts must be bonded together to create a Faraday cage effect and prevent any potential differences between exposed and extraneous conductive parts in the event of a single fault.

If this type of protection is to be used on a TN system additional measures must be applied. The reason for this is that if there were a fault between the neutral and an exposed conductive part, then an earth would be imported into the location via the earthed neutral. This, of course, would defeat the object. Converting this section of the installation to a TT system would be one option.

An additional requirement of this type of protection is that a warning notice compliant with regulation 514.14.1 is fixed in a prominent position at every access point where it is used. The notice must have the following wording:

> The protective bonding conductors associated with the electrical installation in this location MUST NOT BE CONNECTED TO EARTH.
>
> Equipment having exposed conductive parts connected to earth must not be brought into this location.

CHAPTER 42

Protection against thermal effects

This chapter deals with the requirements for protection of persons, property and livestock against the effects of heat or thermal radiation that could be caused by electrical equipment, as well as any fires which may be caused by ignition or degradation of materials, along with protection from smoke and flames that could be caused by an electrical fire in other parts of an electrical installation. The failure of safety services is also dealt with in this chapter.

The requirements for protection against overcurrent are dealt with in Part 4 of this book, which explains chapter 43, and for that reason are not dealt with here.

Although chapter 42 states the requirement of BS 7671 with regard to thermal effects, it must be remembered that there may be some statutory requirements for some installations, particularly with regard to evacuation.

Protection against fire caused by electrical equipment (421) and locations with risks of fire due to the nature of processed or stored materials (422)

It is vital in any electrical installation that electrical equipment does not present a fire hazard to materials that are near to it. Many parts of the installation could produce heat. We are all aware that something as common as a filament lamp could produce enough heat to cause serious damage to surrounding materials. For that reason we take precautions to ensure that the lamp is sited far enough away from anything which could be damaged.

It is not only the heat from the filament lamp which could cause us a problem as often the control gear within luminaires will generate enough heat to cause damage to surrounding materials or worse still ignition or combustion of those materials. Before installing any kind of light fitting, the manufacturer's instructions should be read thoroughly. There may be a symbol on the fitting, in which case reference should be made to table 55.2 of BS 7671 with regard to what the symbol represents. For example, luminaires with the symbol ▽D are designed to have a low surface temperature and those with the symbol ▽F are designed to be fixed to a normally flammable surface. It would be safe to fix luminaires with these symbols to a combustible surface. In many of these areas, there is likely to be quite a lot of dust due to the processes taking place. For this reason, it is recommended that lamps fitted in these areas should also have a degree of protection to IP5X (dust proof).

Of course, it is not only lamps which can get hot; great care and consideration should be given to anything which may produce heat in its normal day-to-day use. Where equipment which is liable to produce heat is to be installed on or near a combustible surface, various methods of protection can be used. For instance, a halogen lamp which would produce a considerable amount of heat should be mounted with the lamp directed away from any adjacent surface and fixed using the correct bracket. This will allow the heat from the lamp to dissipate without causing damage to the surrounding materials.

Any equipment which could produce heat should be mounted on supports or inside an enclosure which will not conduct heat to a degree that could cause harmful thermal effects. Equipment could also be screened by materials with low heat conductance. This could be as simple as glass wool or plasterboard depending on the temperature being generated, and it is also possible to use an air gap, but this would depend on the distances involved. Care should be taken to ensure that any materials used to prevent the transfer of heat do not prevent the dissipation of it; for example downlighters installed in a ceiling should not be covered by thermal insulation (Figures 4.11 and 4.12).

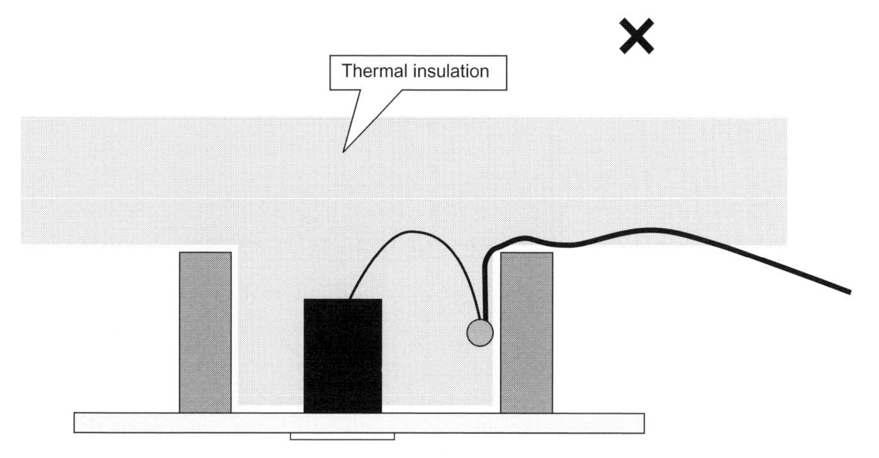

FIGURE 4.11 Bad practice

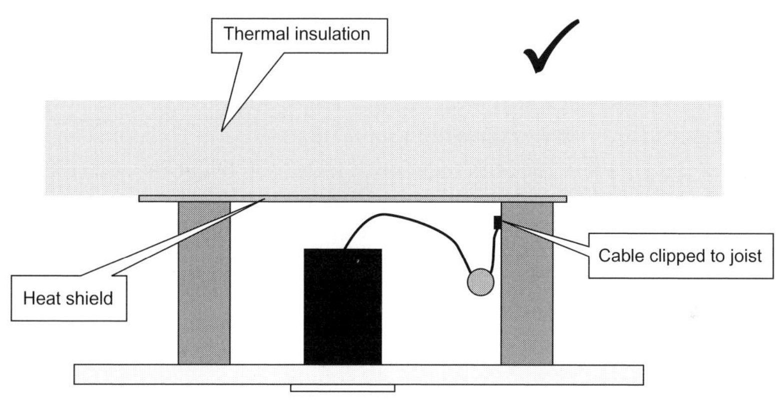

FIGURE 4.12 Correct installation

In some instances, parts of the fixed installation equipment such as contactors, switches and protective devices could produce arcing and sparks at a high temperature. Usually, if we install equipment to the required British Standard, and the equipment is mounted at a safe distance from materials which could be damaged or ignited, we will comply with BS 7671.

Some installations will have within them items of equipment such as transformers or oil-cooled welding machines which contain a significant amount of flammable liquid. In all cases precautions should be taken to prevent the escape of the liquid; however, where the equipment contains more than 25 litres of liquid further precautions are required. For equipment sited outdoors, a simple retention pit filled with pebbles or gravel to prevent the spread of liquid is suitable (Figure 4.13).

For equipment sited in a plant room, further precautions are required such as a fire door and blast-proof walls, along with a retention or drainage pit.

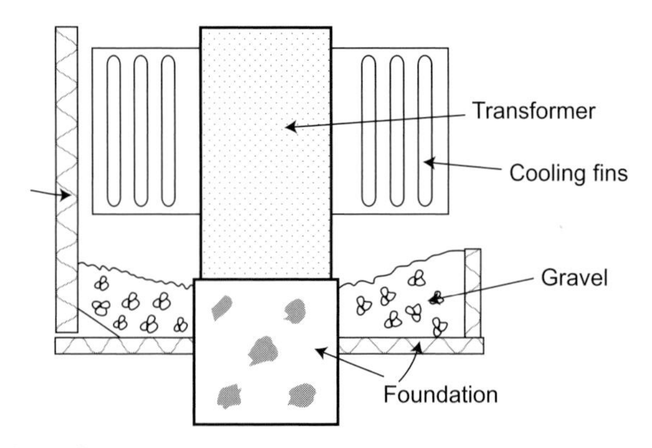

FIGURE 4.13 **Oil-filled transformer**

Precautions where particular risks of fire exist

Consideration must be given to escape routes in buildings which contain materials which could assist the spread of fire. These materials could be stored combustible materials, or materials which have been used in the construction of the building, including the materials used within the electrical installation.

To reduce the risk of flame propagation or the production of smoke and toxic fumes, wherever possible only the electrical equipment which is required for use in the location should be installed. Wiring systems should not pass through these types of location in order to reach another location.

Of course, this is sometimes impossible to avoid because of the way the building is constructed or even due to increased circuit lengths.

In the event of having to run parts of the electrical installation through escape routes, or locations with increased risks of fire due to the nature of stored materials, BS 7671 requires us to ensure that any cables used must be chased in the wall and covered by plaster or concrete. If for some reason the cable has to be run on the surface (clipped direct), then it has to comply with BS EN 50266 or BS EN 60332-1-2. The flat twin and earth 70°C thermoplastic PVC cables which we would normally use (6242Y) are not suitable for use in this type of environment, although 6242B, which looks the same but has XLPE insulation, would be suitable. Firetuff and mineral-insulated cables would also be acceptable. Where a cable has to pass through the location, the cabling should have no connections unless they are contained within a fire-rated enclosure. (Regulation 422.3.4.)

Where the installation is to be in conduit, trunking or cable tray, it would be perfectly safe to use any proprietary brand as these will all meet the requirements as set out by BS 7671. It is also a requirement where the wiring system passes through this type of location, but does not supply equipment within the location, that any connection installed within an enclosure would not burn easily or produce a lot of smoke.

All motors which are not continually supervised must have protection against excessive temperature and any equipment which contains a thermal cut-out must only be manually resettable if the temperature protection operates.

RCD protection must be provided for all circuits which are not wired in mineral-insulated cable or use bus bar or powertrack wiring systems. The operating current of the RCD ($I_{\Delta n}$) must not be greater than 300 mA and where the installation contains heating elements which could cause a fire if a fault were to occur, then the value of $I_{\Delta n}$ must be reduced to 30 mA. Overload protection must also be provided for all circuits which supply equipment in these locations or pass through the location and all circuits must have double pole isolation. This isolation can be achieved by the use of the main switch at the distribution board and need not be for each individual circuit provided that the operational requirements of the installation allow it.

Overload and protective devices and isolators must be installed outside the location, and the neutral as well as the line conductors must always be isolated.

Where a building is constructed of combustible materials or could be classed as a fire-propagating structure due to its shape or design, all of the conditions mentioned earlier in this chapter must be met, along with the possible installation of a fire-detection system. In some cases, this may be linked to fire dampers or shutters placed in ducts or trunking to prevent the spread of fire and smoke. All parts of the structure through which the wiring system passes and which could be deemed a fire barrier, such as partition walls and ceilings, must be reinstated to the same level of fire protection as the material through which it passes. This is true for any fire barrier, not only those in special areas. In some instances, such as where downlighters are fitted, the fire protection could be as simple as fitting fire-rated luminaires or fire hoods.

Clearly, fire protection is of great importance in all areas, but the 17th Edition Wiring Regulations also specifically mention **installations in locations of national, commercial, industrial or public significance**.

These are buildings which could be of importance because of what is kept in them, such as museums and data storage facilities, as well as areas which have a lot of people passing through them, often in confined spaces such as airports, railway stations and national monuments. Evacuation could be a major problem in these areas and for that reason all of the requirements mentioned earlier must be met. This set of regulations also recommends that consideration should be given to provide some additional measures, such as:

- The use of mineral-insulated cables
- Other types of fire-resistant cables
- Cables buried in non-combustible walls, ceilings and floors
- Cables installed in ½-hour fire-resisting partitions or in some cases 1½-hour fire-resistant enclosures, particularly on escape staircases.

Sometimes these methods are not practical due to the age or construction of the building. In these cases, fire safety engineering could be used. Fire safety engineering looks at both structural and safety features; these include fire suppression, smoke control, fire resistance and fire detection.

TABLE 4.4 Maximum permissible surface temperatures

Accessible part	Material of accessible surfaces	Maximum temperature (°C)
A hand-held part	Metallic	55
	Non-metallic	65
A part intended to be touched but not hand held	Metallic	70
	Non-metallic	80
A part which need not be touched for normal operation	Metallic	80
	Non-metallic	90

The regulations only ask that consideration should be given to using these additional measures. It is not a requirement that they are used; however, fire is always a major danger and we should always be prepared to take every possible measure of protection.

Of course, it is not always down to the electrician or the designer as the client may be working to a budget which must allow compliance with the regulations but may not be sufficient to provide the best possible protection. If the requirements of 422.1 are met, then compliance with the regulations will be achieved.

Other issues with heat are the surface temperatures of accessible parts of the fixed electrical equipment which forms part of the installation. Table 4.4 gives the maximum permissible surface temperature of this equipment which is in arm's reach.

CHAPTER 43

Protection against overcurrent

Compliance with the regulations will require that each circuit is protected against overcurrent by using one or more devices which interrupt the supply. Overcurrent could be a current either due to overload or caused by a fault.

Overload current would, in most cases, be due to a circuit carrying more current than it was designed for because too much load has been applied to it. This could be due to more current-using equipment being connected to the circuit than it was designed for, or a piece of equipment developing a fault, such as a motor having to work too hard because of a fault appearing on the equipment which it is driving, or even something as simple as the motor bearings wearing out and seizing up.

In these cases, because the overcurrent is in the circuit due to a problem with the load and not the actual circuit conductors, the current would normally increase slowly. This means that the overcurrent protection will also react and operate slowly.

Fault current is usually due to a fault between live conductors or live conductors to earth, and could be caused by an insulation fault or a mechanical fault within a piece of installed equipment. If the fault is between live conductors it is called a short circuit, and if it is between live conductors and earth it is called an earth fault. In this type of fault, the rise in current is usually very rapid, and the overcurrent protection would be required to react almost instantly, in most cases.

Prospective fault current

The maximum fault current which could flow between live conductors is called PSCC.

The maximum fault current which could flow between live conductors and earth is called PEFC.

We are required to determine the value of each of these two currents, and this can be accomplished by measurement, enquiry or calculation. The highest value must be recorded on any test documents and is known as PFC.

Measurement is usually the best option as it will provide accurate values. Where measurement is not possible, for instance at the design stage of the installation, then enquiry is the only possible method. Regulation 28 of the ESQCR requires that the service provider will give you this information without cost. Calculation is not a practical option in most cases, although it is possible.

Installed equipment in an electrical installation must be capable of withstanding the maximum fault current (PFC) which could flow at the point at which it is installed. Ensuring equipment is to the required British Standard will usually ensure compliance, particularly in domestic installations which are supplied through BS 1361 type 2, or BS 88 fuses which are not rated higher than 100 A. It would be unusual for a domestic installation to have a PFC in excess of 6 kA.

However, fuses and circuit breakers require careful selection in installations other than domestic as these are the parts of the installation which are required to interrupt the supply in the event of a fault.

Protection against overcurrent (fault current or overload) can be provided by using fuses or circuit breakers.

Let us look at fuses first.

Each type of fuse has different characteristics.

Where protection is for *fault current* all types must be able to operate within 0.4 or 5 s depending on the requirements of the circuit or installation which they are being used to protect.

The symbol used to represent the current which will cause the automatic operation of a protective device in the required time due to a fault is known as I_a. For fuses, this is the current which will cause the device to operate within 0.4 or 5 s, depending on the circuit requirements. Circuit breakers will be required to operate at 0.1 s regardless of the type.

BS 3036 semi-enclosed rewirable fuses are perfectly suitable to use provided they are installed correctly, although regulation 533.1.1.3 does say that where a fuse is to be used it should preferably be a cartridge type.

Depending on the device designation, a rewirable fuse will have a rated short circuit capacity of between 1 and 4 kA the service designations are S1A (1 kA), S2A

(2 kA) and S4A (4 kA). The service designation is usually difficult to identify as it is rarely marked on the fuse.

There are two types of cartridge fuse to BS 1361: type 1 has a rated short circuit capacity (I_{cn}) of 16.5 kA, with a type 2 having a rated short circuit capacity of 33 kA. Disconnection times for this type of protective device can be found by using the tables in appendix 3 of BS 7671. Cartridge fuses to BS 88-2 have a rated short circuit current of 50 kA at 415 V, with BS 88-6 having a short circuit rating of 16.5 kA at 240 V and 80 kA at 415 V.

Circuit breakers can be obtained in many rated short circuit capacities. The greater the rated short circuit value the more expensive they will be and it is cost-effective to select the correct rating rather than go for the highest possible.

BS EN 60898 and BS EN 61009 protective devices usually have two short circuit capacities marked on them.

I_{cn}, which is the rated short circuit capacity, is the maximum current which the device could interrupt without causing any damage to other equipment surrounding it.

I_{cs}, which is the in-service rated short circuit capacity, is the maximum fault current which the device could interrupt safely whilst still remaining safe to use. Any current above this value would usually render the circuit breaker unserviceable. As you can see, it is better to select a device with an I_{cs} which is higher than the PFC as the device will still be fine for continued use; however, it is a requirement that the current is not greater than the I_{cn} whether or not the circuit breaker can be reset.

The I_{cs} is usually shown in a rectangle marked on the front of the device. 6000

Unlike fuses which operate due to temperature increase for fault or overload protection, circuit breakers react to fault currents by using a magnetic mechanism within the device and they must interrupt the fault current in a time of 0.1 s or less (Figure 4.14).

BS 7671 only provides us with information which is relevant to equipment which we could purchase and fit at the time it was published. Therefore, the regulations only deal with BS EN 60898 circuit breakers and BS EN 61009-1 RCBOs. To enable these types of protective device to be used on circuits with different types or levels of overload, they are manufactured in three different types – B, C and D. A is not used as identification as it could be confused with the current rating of the device.

We have already seen that the magnetic device within circuit breakers must operate at 0.1 s or less when a fault occurs, either between live conductors and earth or between live conductors. However, as with most things it is not quite that simple, so we need to look at them in a little more detail.

A type B must operate instantly at between 3 and 5 times its rating, a type C at between 5 and 10 times its rating and a type D at between 10 and 20 times its rating. A quick look at figures 22/23 and 24 or the tables in appendix of the regulations will confirm this, although the charts in BS 7671 only show the highest value. For instance, for a 20-A type B they show 100 A for instant disconnection, although in

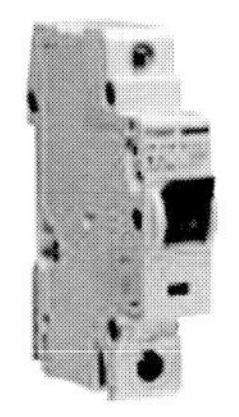

BS EN 60898 is the most onerous Standard written for the manufacturing and testing of circuit breakers. By definition its title 'Specification for circuit breakers for overcurrent protection for household and similar installations', implies that the circuit breaker will be generally used on final circuits. The scope of the Standard states that the circuit breakers are designed for use by uninstructed people and are not designed to be maintained. This is further emphasised by the many tests required to ensure the safety of the person who will return the circuit breaker to service. It does not mean that this Standard only refers to circuit breakers in domestic situations. All buildings have final circuits, and in many instances the first person attempting to reset a circuit breaker could be classed as uninstructed. The Standard recognises this by including the words 'and similar installations' in its title.

Part 2 of BS 7671 defines an instructed person as
'a person adequately advised or supervised by skilled persons to enable him/her to avoid dangers which electricity may create'.
From this an assumption can be made of an uninstructed person.

The scope of BS EN 60898 allows the circuit breaker to be manufactured up to 125A, have a rated voltage of 440V and a breaking capacity of up to 25kA. The inclusion of an energy let-through class rating provides the designer with a better indication of the ability of the device to operate under fault conditions.

Information about the device can be found on the label and also in the manufacturer's literature.

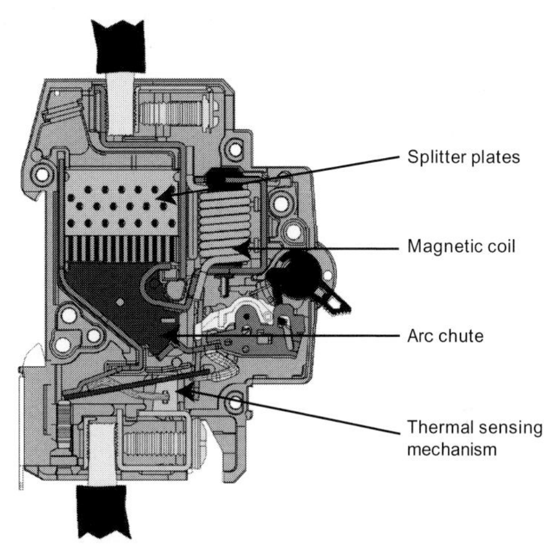

FIGURE 4.14 **BS EN 60898 circuit breakers**

reality it could be anything between 60 and 100 A. The reason for the different types is to prevent circuits such as motor or transformer circuits nuisance tripping these devices due to the high levels of current drawn on the initial startup of the motor or transformer, or indeed anything else which may have high startup (inrush) currents such as fluorescent lighting (Figures 4.15, 4.16 and 4.17).

Overload

A protective device is installed in a circuit to prevent damage to cables due to overcurrent. This could be fault current whereby the device has to operate very quickly due to the magnitude of the current. It could also be overload current, which could be due to faulty equipment or even too great a load being connected to the circuit.

Where protection is required for overload, we have different requirements which we must comply with. As stated in an earlier section, overload current usually builds up slowly due to something other than an installation fault. Our protective devices have to be able to interrupt the supply when the overload is likely to be harmful to the circuit cables, due to the increase in conductor operating temperature.

Devices used for overload protection have to react more slowly than when they are being used for fault protection.

Rewirable fuses need to be carefully selected when being used for overload protection as they are not very accurate, and will operate on an overload of

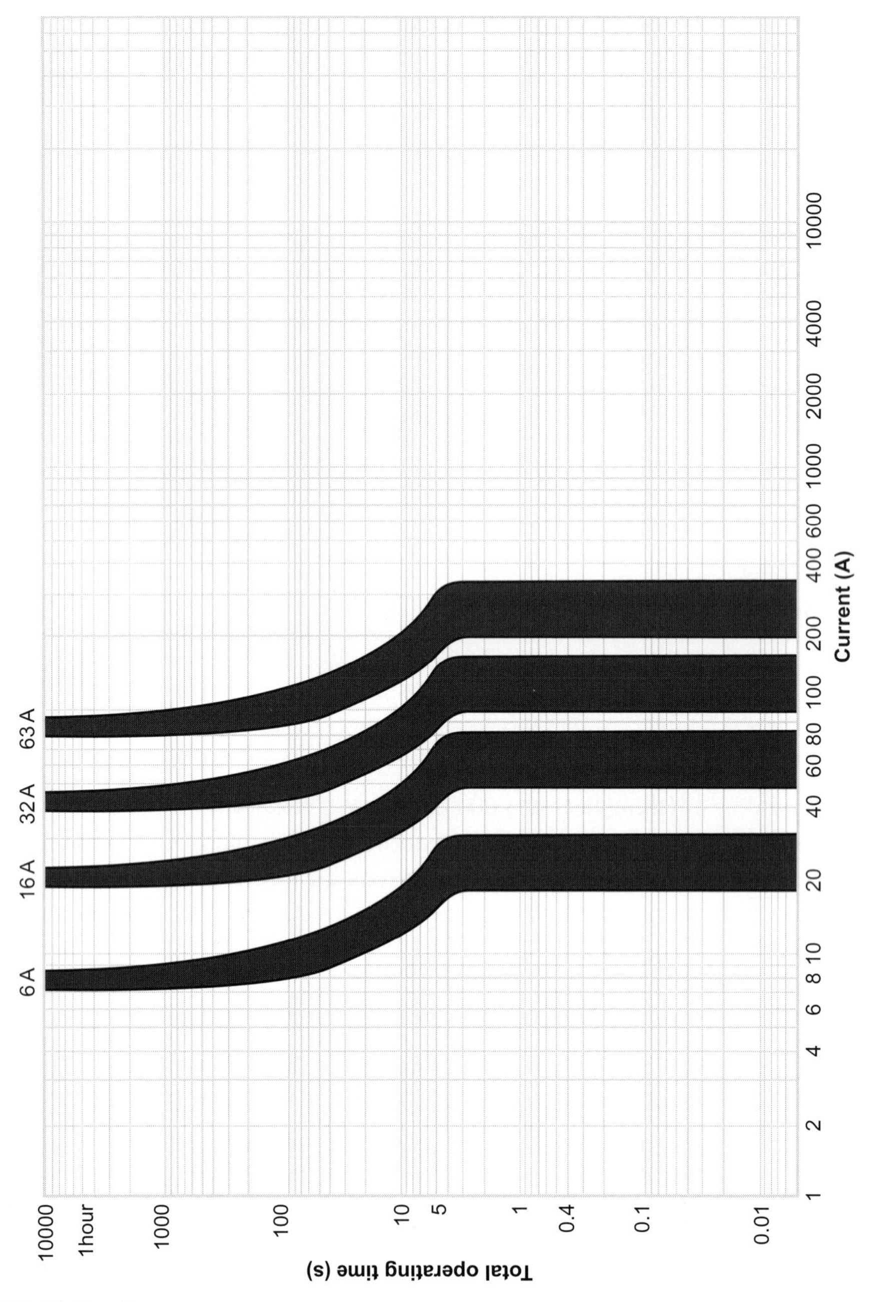

FIGURE 4.15 **Type B**

twice their current rating in around 4 hours, which will not satisfy the requirements of BS 7671.

Regulation 433.1.1 tells us that a protective device must operate on an overload of 1.45 times the rating of the cable which it is protecting. It is also required to operate within 1 hour.

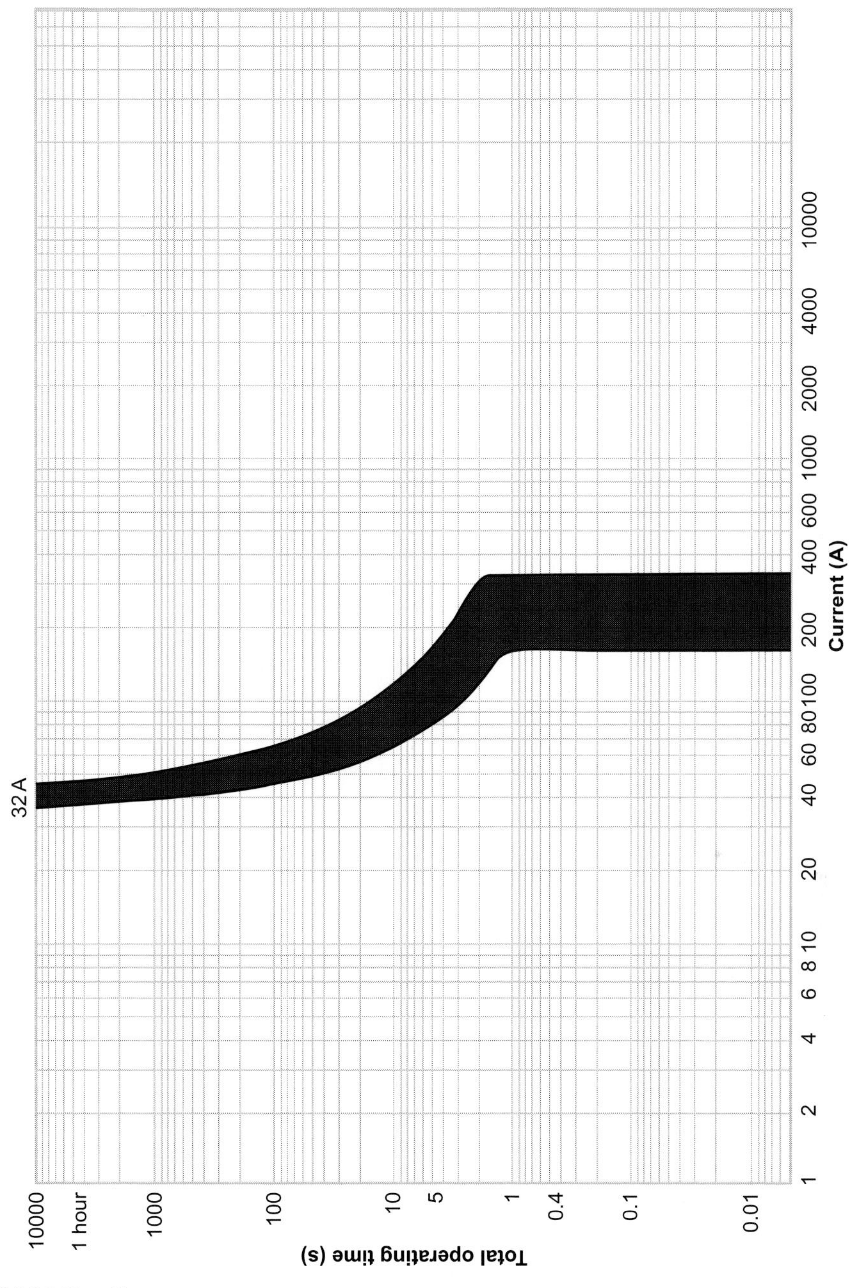

FIGURE 4.16 Type C

If we assume that we have a load with a design current of 20 A and that we are to use a 20-A BS 3036 fuse for protection, it would appear to satisfy the regulations as the design current (I_b) of the circuit is not greater than the current rating of the protective device (I_n) $I_b \leq I_n$ and we could use a cable with a rating of 20 A for the circuit.

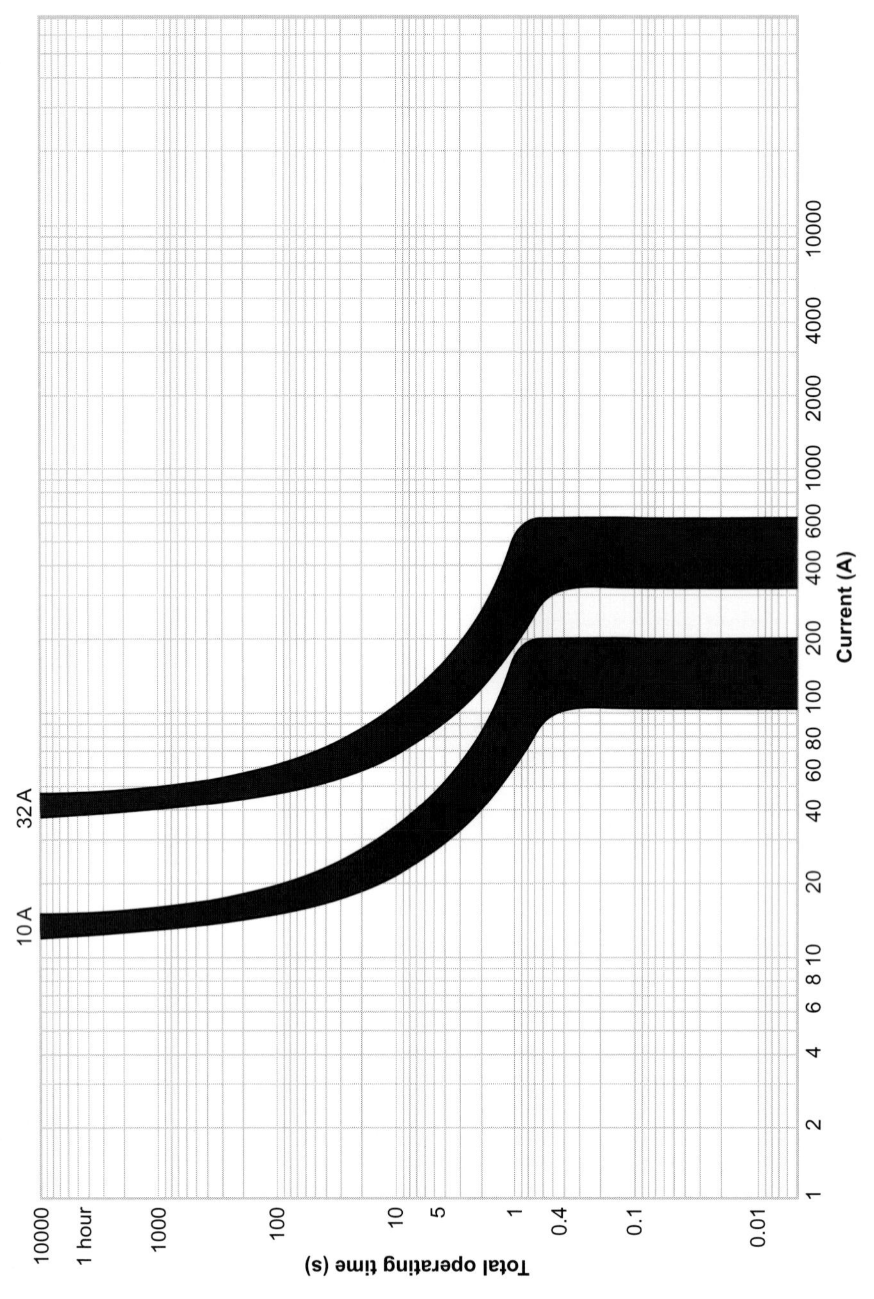

FIGURE 4.17 **Type D**

Of course, it will not comply because the device will not operate until a current of at least 40 A is reached, which is twice the rating of the cable conductor.

To compensate for this, we need to install a larger conductor that will cope with the heat in the cable due to the overload. To calculate the correct size of cable to compensate for the additional current we have to use a factor for a rewirable fuse.

We know that the device will not operate until it has reached twice its rating. We also know that we require it to operate at 1.45 times the rating of the cable.

If we divide 1.45/2, it will give us the ratio of the two figures: 0.725.

We are now required to use this figure as a factor which we must divide into the rating of the protective device. This will calculate the minimum current which the conductor must be able to carry.

The rating of the protective device is 20 A so the calculation is: 20/0.725 = 27.58. This is the minimum value of current that the cable should be rated at.

As an explanation of how this works, we know that the 20-A protective device will not operate until a current of 40 A is reached. We also know that the protective device must operate at 1.45 times the rating of the cable.

Our calculation shows that we must use a cable with a minimum rating of 27.58 A. If we multiply 27.58 × 1.45, we have 40 A. It is no coincidence that 40 A is the current which will operate the fuse.

Regulation 433.1.3 requires that where overload could occur and the protective device is a BS 3036 fuse, we must always use a factor of 0.725 to ensure that the cable will not be greatly overloaded. As we have seen from the earlier exercise, the nominal current of the device can be divided by the 0.725 factor to ensure that we select a larger cable.

We can approach the problem in another way if it suits us better. Sometimes we may need to know what value of current a conductor could carry under particular installation conditions. This value is known as I_z and it is calculated as follows.

Let us assume that we have a circuit wired in 4 mm² 70°C twin and earth cable which is protected by a 20-A BS 3036 rewirable fuse. It is enclosed in conduit in an insulated wall and the conduit contains one other circuit. The ambient temperature will not be above 30°C.

We need to know what current this cable can carry under these installation conditions (I_z).

The first part of the exercise is to reference any rating factors which may apply to this cable. We know that when a BS 3036 fuse is used then 0.725 is the factor (C_c).

The circuit is in a conduit with another circuit. This will require us to find the factor for grouping (C_a) from table 4C1 in appendix 4. We can see that two circuits enclosed or embedded have a rating factor of 0.80. Reference to table 4B1 will give us a rating factor of 1 for our cable at 30°C.

Using these rating factors, we can calculate the current which the cable will be able to carry safely without causing it any damage due to heat from overload. Table 4D5 in appendix 4 will give us the current at which the cable is rated before any factors are applied (I_t); we can see that from column 7, a 4-mm² conductor has an I_t of 26 A.

We must now multiply I_t by the rating factors to find I_z.

$$I_z = 26 \times 0.725 \times 0.8 \times 1 = 15.08\,\text{A}$$

This is the current which the 4-mm² cable can carry under the known conditions.

Cartridge fuses

These protective devices operate on a rise of temperature in the same way as a rewirable fuse. The difference is that these fuses are manufactured to precise values using metals such as silver for the fuse link, unlike the rewirable fuse which is just a piece of copper wire between two terminals.

For this reason, BS 88 and BS 1361 fuses are very accurate and will comply with the requirements of BS 7671 with regard to overload and fault current.

Cartridge fuses can be obtained with different characteristics for different types of circuit – they can be for general use, motor circuits, etc. It must be remembered that BS 88 fuse carriers and links should not be used in domestic consumer units as the physical size of the fuse does not alter with the nominal fuse rating. BS 1361 carriers are designed specifically for use by instructed personnel and are therefore suitable for this type of installation.

Circuit breakers

BS EN 60898 circuit breakers and BS EN 61009-1 RCBOs also have a product standard which requires the thermal trip part of the device to operate within specific times when an overload occurs, unlike the fault current part of the device which has different requirements for a type B, C and D. The overload part of the device has the same characteristics for all types.

The current which causes the device to trip in the conventional time is known as I_2 and the requirements are:

- No trip up to 1.13 times the nominal rating (I_n) of the device.
- Must trip within 1 hour at between 1.13 and 1.45 times the nominal rating (I_n) of the device.

(For devices above 60 A the time is 2 hours.)

At 2.55 times the nominal rating (I_n) circuit breakers up to 32 A must operate within 1 minute any circuit breaker above 32 A and must operate within 2 minutes.

Regulation 433.1.4 requires that any cable which is buried directly in the ground is derated by 10%. This is because the heat from the cable will not dissipate very well as the soil around the cable will also heat up. By derating the cable by 10%, we ensure that the current flowing through the conductor will not increase the cable temperature to its maximum permissible under normal circumstances.

433.2 Position of devices for protection against overload

If we wanted to wire a lighting point in a loft, we would normally connect into the nearest lighting point and pick up a supply from there, which in most cases is the sensible and easiest option. There are instances, however, where we need to obtain a supply from a circuit which is rated higher than the supply which we need. For example, if we wanted to wire a light point in a loft and the only accessible circuit was a 4-mm^2 radial most of us would cut a joint box into the 4-mm^2 radial and come from that joint box to a fused connection unit, from where we would run a 1-mm^2 to the lighting point.

Regulation 433.2.2 allows us to connect into the joint box using the 1-mm^2 which would be required for the light and fit the fused connection unit up to 3 m away provided we install the cable in such a way that we reduce the risk of faults and fire. In other words, as long as we install the cable neatly and out of harm's way we will comply. This regulation applies to any circuit which requires the reduction of cable size and is sometimes very useful.

433.3.1 Omission of devices for protection against overload

This is a very useful regulation although it is not very often taken advantage of. Overload protection need only be provided for circuits that will possibly overload; this could be a ring circuit or radial where the installer of the circuit has no control over what is being connected to the circuit.

Where the load which is connected to the circuit cannot cause an overload, such as most water heaters (storage or instantaneous), or lighting circuits with fixed loads, such as discharge lighting, overload protection is not required unless the circuit is in a location which presents an increased fire risk. In many situations, this will allow us to use smaller cables.

Let us take for example a 9.5-kW electric shower wired in 70°C PVC twin and earth cable which is enclosed in conduit in an insulated wall and protected by a BS EN 60898 circuit breaker.

The design current (I_b) would be 9500/230 = 41.3 A.

Under the usual circumstances, we would now be required to select a protective device (I_n) which is equal to or greater than the load. The nearest off-the-shelf device would be 50 A. We would then divide I_n by any rating factors affecting the cable.

$$I_t \geq \frac{I_n}{\text{Rating factors}}$$

This would then give us the required value I_t of the cable. If we now refer to table 4D5 column 7, we can see that even if there were no rating factors this would result in installation of a cable of at least 16 mm^2.

If we refer to regulation 433.1.1, we will be able to use the following formula:

$$I_t \geq \frac{I_b}{\text{Rating factors}}$$

This can be found in appendix 4 of BS 7671, section 5.2. To comply with this formula, all we have to do for our example is to select a cable which can carry the load current, which is 41.3 A. This is because the shower cannot overload: it will either draw 41.3 A or if it goes open circuit nothing at all, unless it goes short circuit in which case the protection will be fault protection and will be provided for by the 50-A device, which will be the nearest standard device which is rated above the load current.

Another example of this would be an immersion heater which is rated at 3 kW, wired in 70°C twin and earth cable which is in a stud wall touching the inner wall surface and protected by a BS 3036 fuse.

The conventional calculation would be $I_t \geq I_n/C_c$ or 15/0.725. This would result in a cable with a minimum rating of 20.68 A being required. From table 4D5 column 2, we can see that a 2.5-mm^2 cable would be required.

Regulation 433.1.1 allows us to use $I_t \geq I_b$/rating factors.

Note 1 of this regulation leads us to 433.1.3 which tells us that the fuse rating factor C_f is only required to be used when the fuse is protecting against overload. Section 5.2 of appendix 4 also confirms this.

Therefore, in this case if we look in table 4D5 column 2, we can see that a 1-mm^2 cable could be used. Clearly, this sort of calculation needs to be used carefully and voltage drop along with Z_s values have to be taken into account.

Let us say that our circuit is 20 m long and the Z_e of the system is 0.8 Ω.

From appendix 12 BS 7671, we can see that we are allowed a maximum of 5% voltage drop in this circuit, which is:

$$\frac{230 \times 5}{100} = 11.5\,\text{V}$$

If we now look in at table 4D5 column 8, we can see that the voltage drop for a 1-mm^2 copper cable is 44 mV/A/m. This circuit will have a voltage drop of:

$$\frac{44 \times 13 \times 20}{1000} = 11.44\,\text{V}$$

As we can see, this is satisfactory, so we can now calculate Z_s. From the example, we know that this circuit is protected by a 15-A BS 3036 fuse. As the rating of the fuse is less than 32 A it must operate in 0.4 s or less. The maximum Z_s for this circuit is found in table 41.2 and we can see that it is 2.55 Ω.

We now need to look in table 9A of the on-site guide or table 1A of guidance note 3 to find the $R_1 + R_2$ value of a 1-mm^2 live conductor with a 1-mm^2 CPC. It is 36.2 mΩ/m at 20°C. The calculation is:

$$\text{Actual } Z_s = Z_e + R_1 + R_2$$

We know that $Z_e = 0.8\,\Omega$,

$$R_1 + R_2 \text{ for the cable } = \frac{36.2 \times 20}{1000} = 0.72\,\Omega$$

This is the value at 20°C. We need to know what it will be when it reaches its maximum operating temperature of 70°C. The resistance of copper conductors will increase by 2% for each 5°C rise in temperature, so as the conductors will rise in temperature by 50°C the resistance will increase by 20%. We need to use a multiplier of 1.2, which will increase the value of resistance by 20%. This multiplier can also be referenced in tables 9c of the on-site guide and 1c of guidance note 3.

$$0.72\,\Omega \times 1.2 = 0.87\,\Omega$$

Actual Z_s will now be $0.8\,\Omega + 0.87 = 1.67\,\Omega$. This will satisfy the requirements of BS 7671, as it is less than the maximum permissible for the fuse, which is $2.55\,\Omega$.

Of course, most of us use 2.5 mm^2 for the wiring of immersion heater circuits, which is fine as it usually does not require any calculation as we know it works; however, if we comply with the regulations a considerable reduction in cable sizes could be achieved.

Where motor circuits are installed, the protective device for the circuit is only required to provide fault protection. The overload protection will be provided by the motor starter overload device. Similar calculations can be used for these circuits as for other circuits which cannot overload.

433.4 Conductors in parallel

It is not uncommon for conductors to be installed in parallel; the most common use is in ring circuits. This allows us to use smaller conductors and supply socket outlets to a greater floor area through one circuit. This is because the cables are installed in a loop, which is two cables in parallel. This method gives the circuit twice the current carrying capacity of a single cable; it also reduces voltage drop to approximately a quarter of the voltage drop which the same length of cable would have if installed as a radial.

When cables are used as a ring circuit, it is permitted to spur off from the ring with a single cable. Although this practice results in a single 2.5 mm^2 cable being protected by a 30- or 32-A protective device the cable cannot be overloaded because the size of the load which could be supplied by the socket outlet is limited by the outlet ($2 \times 13 = 26$) (Figure 4.18).

In larger installations, the installing of conductors in parallel has other benefits. Large steel wire armoured cables are often very difficult if not impossible to bend to a tight enough radius, particularly when trying to connect into a large distribution board or bus bar chamber, etc. Using two or more cables allows the use of smaller cables which are easier to bend and can be installed with a tighter radius.

Another benefit is cost saving. For example:

A steel wire armoured multicore cable is to be installed on a perforated cable tray and it is required to carry a three-phase load of 440 A. If we look at column 9 in table 4D2A of BS 7671, we can see that we would need to use a 300-mm^2 cable. In this instance, as the regulations allow us to use conductors in parallel, we could use three 50 mm^2 cables. This would give us a current carrying capacity of 153 A per cable, which would be $3 \times 153 = 459$ A. In addition to easier installation of the cables, we would make a considerable financial saving by using only 150 mm^2 cross-sectional area of copper instead of 300 mm^2.

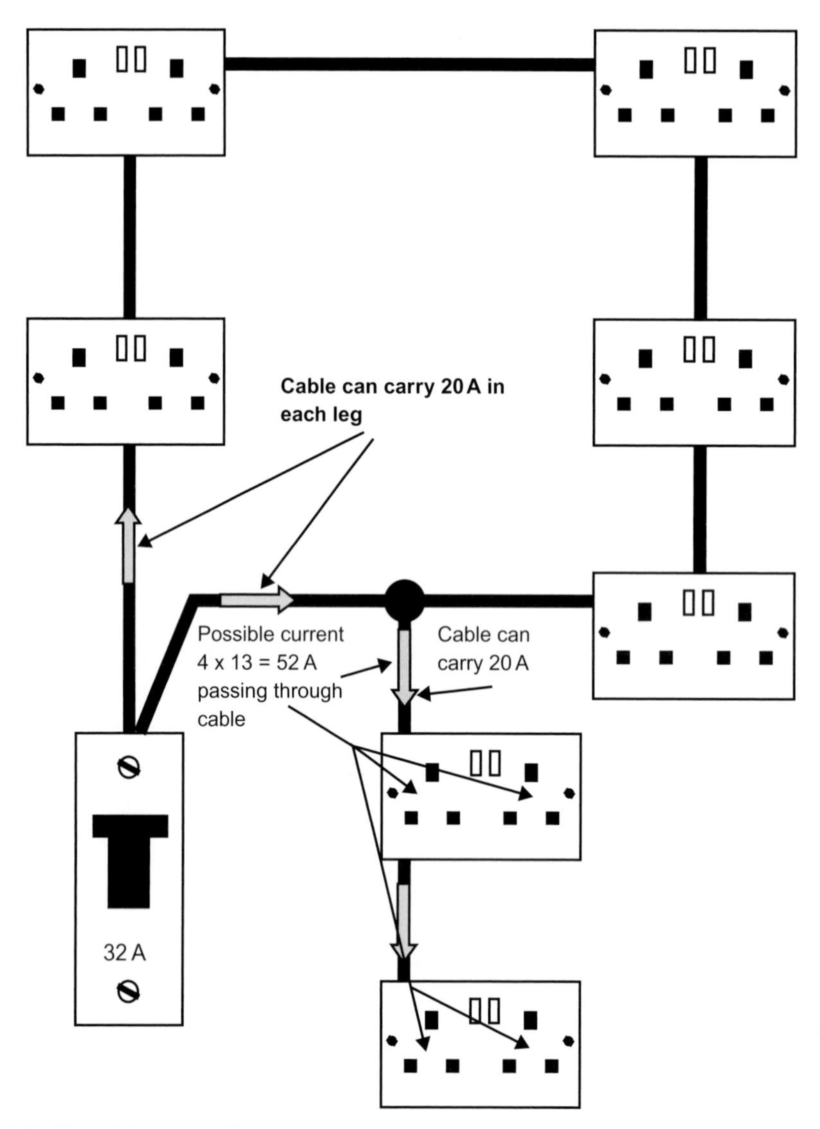

FIGURE 4.18 **Ring with non-compliant spurs**

Appendix 10 of BS 7671 gives more information on the installation of conductors in parallel. In all cases the cables must be of the same cross-sectional area, length and material. Where two multicore cables are in parallel either one protective device can be used or each cable can have its own protective device. In both cases, these should be installed at the supply end of the circuit. Where more than two cables are used, consideration must be given to using a protective device at the supply and load end. This is to prevent the current from flowing back down the conductor with the fault via the other conductors (Figure 4.19).

A better option would be to use either one protective device or linked protective devices.

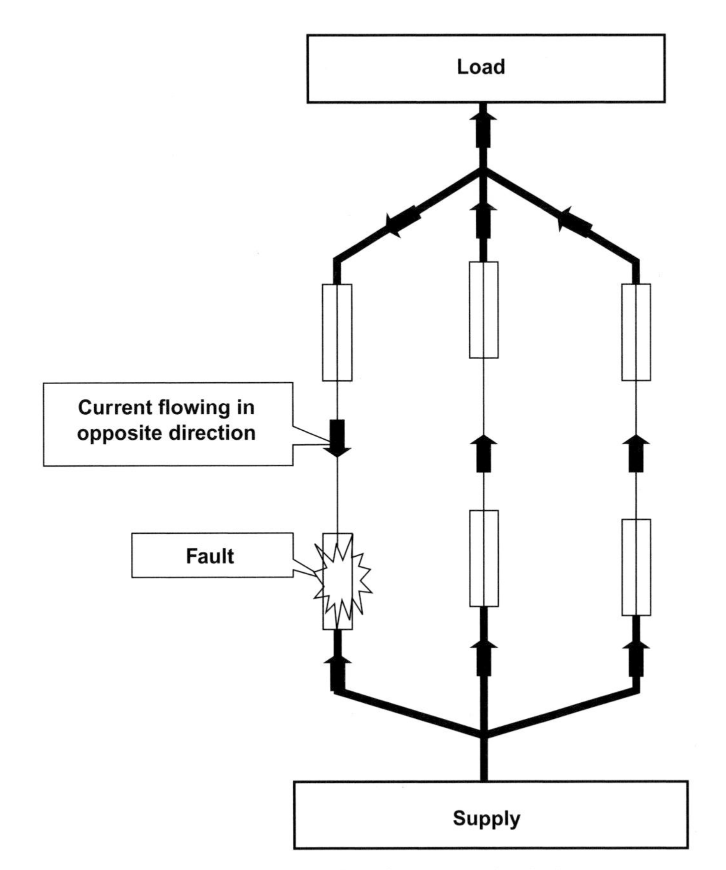

FIGURE 4.19 **Fault remaining after only one operation of one protective device**

Where multicore cables are used equal current sharing between the conductors is easily achieved simply by using the same length, cross-sectional area and conductor material; however, where single core conductors are to be used in parallel, very careful consideration must be given to the configuration of the cables to ensure that they carry an equal amount of current. Further information about this can be found in appendix 10 of BS 7671 or guidance note 6.

434.5 Characteristics of a fault current protective device

Wherever possible it is better to install a device that is capable of breaking the maximum fault current which could flow through it without damage. It must operate quickly enough to clear the fault before damage occurs to the cable.

Use of regulation 434.5.1 allows us to use a device with a lower rated breaking capacity where another protective device is installed upstream which has the necessary short circuit rating, provided that the energy let-through of the upstream device is not greater than the PFC rating of the device which is installed downstream. In some instances, this would allow us to install a BS 3036 rewirable fuse which may only have a PFC rating of 1 kA, provided the protective device upstream had a

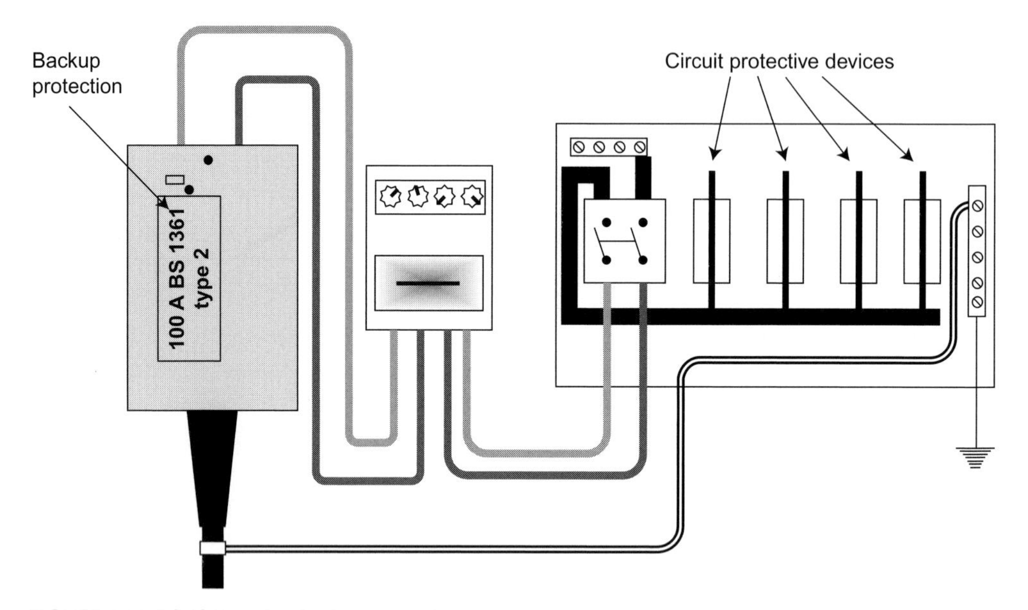

FIGURE 4.20 **BS 1361 used as backup protection**

let-through energy of less than 1 kA. Manufacturer's information with regard to let-through energy of devices should be consulted in these cases.

In domestic installations where the supply has a protective device which is a type 2 BS 1361 with a rating of no greater than 100 A, it is generally accepted that back up protection is provided by the supply protective device (Figure 4.20).

When a fault occurs it is extremely important that the current is interrupted before the temperature of the conductors increases to a level where damage will occur to them or the cable insulation. The temperature of the conductors under fault conditions will depend on the:

- Value of the fault current (I_f)
- Material of the conductor (k)
- Size of the conductor (mm^2)
- Time it takes for the fault current to be interrupted (t seconds).

As an example, let us assume that we have a circuit wired in $10\,mm^2$ copper thermoplastic 70°C singles. The supply has a PFC of 3 kA and the protective device for the circuit is a 50-A BS EN 60898 C type circuit breaker.

If we look at table 43.1 of BS 7671, we can see that the value k for this conductor is 115 as it is below $300\,mm^2$. We now have all of the values apart from the maximum time for which a fault current of 3 kA could be allowed to flow in the circuit before causing damage.

The formula used to calculate the maximum permissible time for which the fault current could flow is:

$$\frac{k^2 s^2}{I^2} = \mathrm{t}$$

If we now add the figures from our example:

$$\frac{115^2 \times 10^2}{3000^2} = 0.146\,\mathrm{s}$$

The maximum time which we could allow the fault to continue for is 0.146 s. If we look at figure 3.5 from appendix 3 of BS 7671, we can see that the device will operate in less than 0.1 s; therefore, no damage will occur to the conductor or the insulation.

435.2 Protection offered by separate devices

In some cases, fault current protection and overload protection are offered by separate devices. A good example of this would be where fault current protection is provided by a BS 88 motor fuse and overload protection is provided by a motor starter.

Compliance with regulation 435.2 requires that we ensure that the motor starter will be able to withstand the let-through energy of the fuse. In most cases, equipment manufactured to the required British Standard will be suitable, but where large fault currents are likely to be present the starter manufacturer should be consulted.

Overcurrent and fault current protection is not required for conductors supplying certain types of equipment which are supplied from a source, such as a bell transformer or welding transformer, which is incapable of supplying currents exceeding the capacity of the conductors supplying the load. (Regulation 436.)

CHAPTER 44

Protection against voltage disturbances and electromagnetic disturbances

This section of the regulations contains four parts, although one of them is simply a number for a section which is reserved for future use.

442

This is a new section in BS 7671 which deals with overvoltages due to earth faults in the substation. Knowledge of this section will not be required by electricians who are working on low-voltage installations, unless the installation includes a privately owned substation.

443

Lightning strikes and the switching of equipment within an installation can cause voltage surges which could cause damage to an electrical installation and the equipment connected to it.

Where an installation is installed in an area where there are more than 25 thunderstorm days a year, overvoltage protection (surge protective devices) is required along with the correct selection of impulse withstand voltage for equipment. Fortunately, in the UK we do not have any particular areas which have this number of thunderstorm days. Because of this, we do not need to consider the installation of surge protective devices to enable us to comply with the current edition of BS 7671.

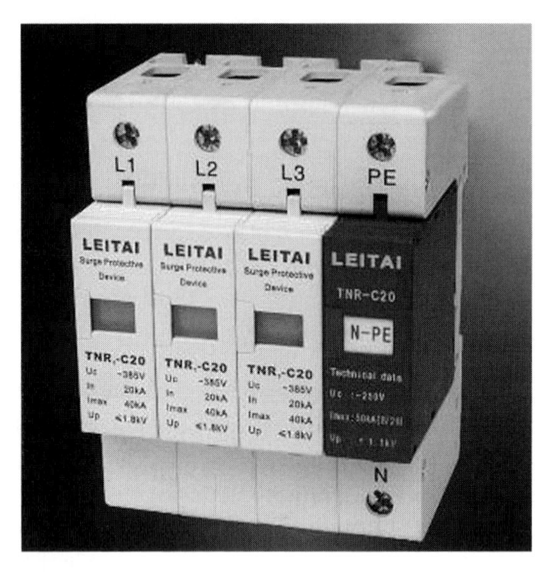

FIGURE 4.21 **Surge protection**

Although it is not a requirement, surge protection may still be considered where damage to valuable equipment could result in overvoltages (Figure 4.21).

In appendix 5 of BS 7671, overvoltages are categorised as:

- AQ1 $\leq$ 25 thunderstorm days per year (applies to the UK)
- AQ2 $>$ 25 thunderstorm days per year
- AQ3 direct exposure.

The requirements of the minimum impulse withstand voltage of equipment are given in table 44.3 and examples can be found in table 44.4.

If we install equipment to a British Standard and in areas where it is intended to be used, we will have complied with these requirements.

445 Protection against undervoltages

The loss of supply or undervoltage could cause severe damage or injury; for example a significant reduction in voltage could result in motors overloading.

Regulation 552.1.3 requires that any motor which is rated at greater than 0.37 kW must have overload protection. Generally, this is a magnetic or thermal device within the motor starter which would interrupt the supply to the starter coil in the event of an overload; this overload could be caused by a reduction in voltage.

A contactor coil will open when the voltage reduction reaches around 20%.

The dangers involved if the motor or the equipment being controlled by the contactor were to start operating when the voltage returned to normal could be life-threatening.

Manual resetting must be provided for any equipment connected to the electrical installation which may cause danger if it is suddenly re-energised. Motor starters and emergency stop systems are common examples of compliance with this requirement.

Part 5

In this Chapter:

SELECTION AND ERECTION OF EQUIPMENT

Chapter 5 consists of six sections and provides us with the requirements for the selection and erection of electrical equipment.

CHAPTER 51

This section provides us with the selection of equipment for compliance with BS 7671.

For compliance we need to ensure all of the equipment which we install is to a British or Harmonised Standard. If it is not we must, be certain that it will not be providing a lesser degree of safety than that which would be required by a British Standard. If you install it, you take responsibility for it, no matter that your customer has said that they will accept responsibility.

Remember you are the skilled responsible person: your customer may have no idea of the possible dangers.

Compliance with a standard will require that the manufacturer's instructions are read and that the equipment is used as intended.

512 Operational conditions and external influences

All installed equipment must be suitable for use with the following.

Voltage

The lowest and highest voltages need to be considered.

A Practical Guide to the 17th Edition of the Wiring Regulations. DOI: 10.1016/B978-0-08-096560-4.00005-9

Current

Equipment must be suitable for the current which it is required to carry. In general, all equipment must be able to operate at a temperature of 70°C.

This is the maximum conductor operating temperature of our general-purpose thermoplastic cables. Some cables can operate at higher temperatures, such as mineral-insulated or thermosetting cables. If cables are to be used which can operate at a temperature of greater than 70°C, it must be confirmed that the equipment to which the conductors are being connected can withstand a higher temperature.

Where cables such as 90°C thermosetting are being used, they must have their current carrying capacity calculated as for a 70°C cable of the same cross-sectional area. As an example, a 2.5-mm^2, two-core 90°C thermosetting cable from table 4E4A will carry a maximum of 36A when clipped direct. If we now look in table 4D4A, we can see that the same size 70°C thermoplastic cable clipped direct can only carry 28A. Obviously, the conductor is of the same size and material; the difference in the rating is due to the different type of insulation.

The British Standard requires that equipment must withstand a maximum temperature of 70°C. If a cable rated at 90°C is to be used, we have two choices: one is to use equipment which the manufacturer confirms is suitable for the higher temperature; the second is to ensure that the conductor does not reach a temperature in excess of 70°C by using tables which give us the maximum current ratings for 70°C cables.

Frequency

All installed equipment must be suitable for the frequency of the supply to which it is connected.

Power

Any installed equipment must be rated to enable it to do the job which it is intended to do. There would be little point in installing an electric motor rated at 1kW to drive a load requiring 1.5kW. Apart from damaging the motor, overload could occur in the circuit.

Compatibility

When equipment is installed, consideration must be given to whether or not it will have a detrimental effect on any other equipment installed within the installation.

Effects which we need to consider include:

- Vibration
- Heat
- Switching operations
- Interference with IT and telecommunication equipments.

Of course, this is not an exhaustive list as there are many other harmful effects which could arise within an electrical installation.

Impulse withstand voltage

If we install equipment to a BS or European Standard, we will have no problem complying with this regulation. We must remember though that any equipment must be installed to the manufacturer's instructions and only used for what it is intended.

External influences

Appendix 5 of BS 7671 provides us with a concise list of external influences. IP ratings are also used to describe protection against the ingress of dust and water (Figure 5.1).

All equipment must be suitable for use where it is to be installed. Sometimes equipment will not be suitable for the location in which it is to be situated; in these cases additional protection may be added during the installation. Where additional protection is added care must be taken to ensure that it does not have a detrimental effect on the equipment; an example of this could be where the equipment has been enclosed and there is a lack of ventilation. Great care must be taken when cables are installed where the temperature is likely to rise above 30°C as any calculation which we carry out is usually based on this temperature. If a rise in temperature is expected, then this must be taken into account at the design stage of a circuit to ensure the correct selection of the conductor size.

513 Accessibility

A major consideration when installing equipment which may at some point in time require maintenance, repair, adjustment or replacement is that it needs to be accessible. There is nothing worse than not being able to get the cover off a piece of equipment because a cupboard has been built around it and the door will not open fully, or somebody has installed a motor in ducting and not left an inspection cover which can be removed to allow access to the motor adjustment. How often have you been asked to change a synchronous motor on a motorised valve which has been put behind a central heating cylinder?

It is possible to make a joint in a cable and bury it in a wall or under a floor provided the joints comply with the requirements set out in section 526. The joints have to be crimped or soldered and also have to be in a suitable enclosure. Clearly, any joint that may work loose has to be accessible. Just think how much easier some jobs would be if when we had to use a joint box, we left a little note next to the distribution board identifying where it was; it is such a simple thing to do but rarely does it get done.

514 Identification and notices

Within an electrical installation, identification is very important. This section of the regulations provides us with a list of items which we should consider.

First numeral Protection against ingress of solid foreign objects			Meaning for protection of persons against access to hazardous paths
IP	**Requirements**	**Example**	
0	No protection		No protection provided
1	Full penetration of 50 mm diameter sphere not allowed and shall have adequate clearance from hazardous parts. Contact with hazardous parts not permitted	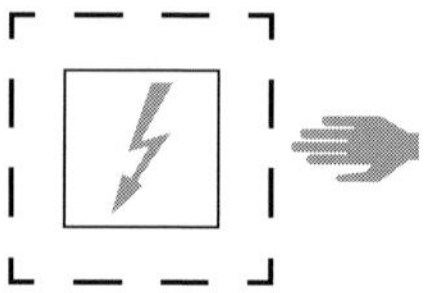	Back of hand
2	Full penetration of 12.5 mm diameter sphere not allowed. The jointed test finger shall have adequate clearance from hazardous parts	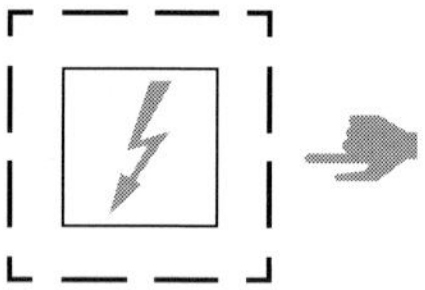	Finger
3	The access probe of 2.5 mm diameter shall not penetrate	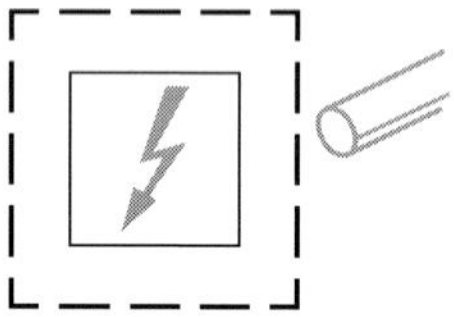	Tool
4	The access probe of 1.0 mm diameter shall not penetrate	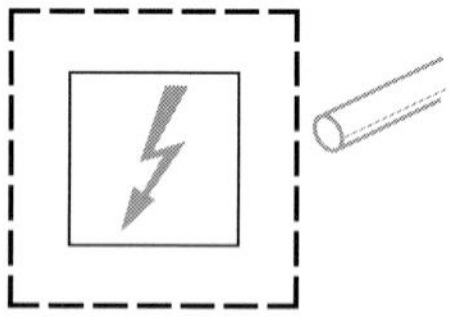	Wire
5	Limited ingress of dust permitted (no harmful deposit)		Wire
6	Totally protected against ingress of dust		Wire

FIGURE 5.1 **(A) IP codes: first number**

Second numeral			
Protection against ingress of water			Meaning for protection from ingress of water
IP	**Requirements**	**Example**	
0	No protection		No protection provided
1	Protection against vertically falling drops of water. Limited ingress permitted		Vertically dripping
2	Protected against vertically falling vertical drops of water with enclosure tilted 15° from the vertical. Limited ingress permitted	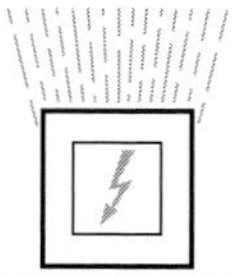	Dripping up to 15° from the vertical
3	Protected against sprays to 60° from the vertical. Limited ingress permitted		Limited spraying
4	Protected against water splashed from all directions. Limited ingress permitted		Splashing from all directions
5	Protected against low pressure jets of water from all directions. Limited ingress permitted	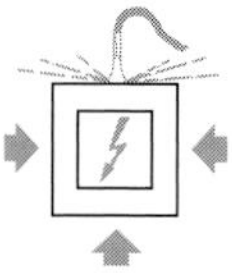	Hosing jets from all directions
6	Protected against strong jets of water		Strong hosing jets from all directions
7	Protected against the effects of immersion between 15 cm and 1 m	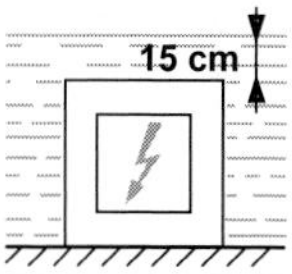 15 cm	Temporary immersion
8	Protected against longer periods of immersion under pressure	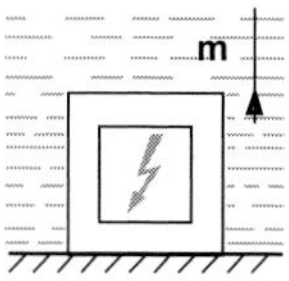 m	Immersion

FIGURE 5.1 (B) IP codes: second number

Additional letter (optional)

IP	Requirements	Example	Meaning for protection of persons against access to hazardous paths
A	For use with first numerals **0** Penetration of 50 mm diameter sphere up to guard face must not contact hazardous parts	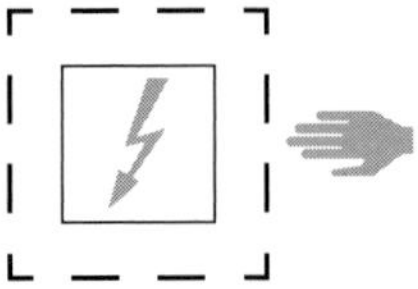	Back of hand
B	For use with first numerals **0 & 1** Test finger penetration to a maximum of 80 mm must not contact hazardous parts	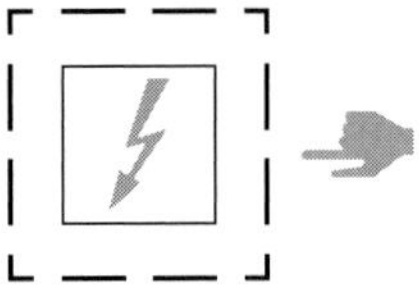	Finger
C	For use with first numerals **0, 1 & 2** Wire of 2.5 mm diameter × 100 mm long must not contact hazardous parts when spherical stop face is partially entered	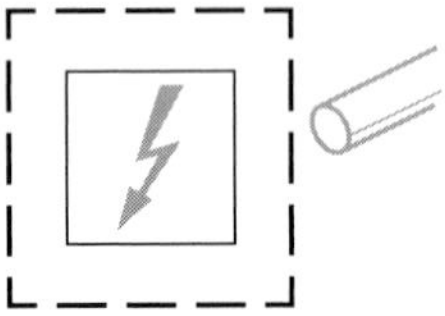	Tool
D	For use with first numerals **0, 1, 2 & 3** Wire of 1.0 mm diameter × 100 mm long must not contact hazardous parts when spherical stop face is partially entered	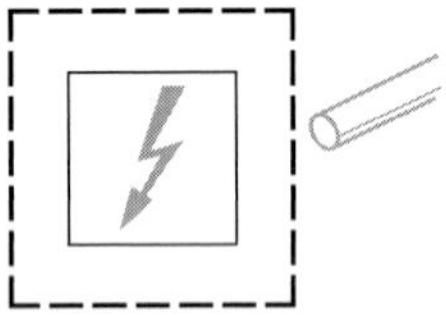	Wire

Limited penetration allowed with all four additional letters

FIGURE 5.1 **(C) IP codes: additional letter (optional)**

Conduit

Where conduit is installed alongside other pipework, and it is not obvious which pipework is which, the conduit should be painted orange. It is not the intention that all conduit is identified in this way.

Identification of conductors

All conductors must be identified; this can be by colour, lettering or numbering, whichever is the most suitable. In an ideal world, a conductor would be identified for its entire length, but this is not always practical. Identification at the point of termination is usually quite acceptable, it is a requirement that the colours comply with BS 3858 and table 51 of BS 7671.

As electricians, we know that the neutral is blue. We also know that great care must be exercised when working on older three-phase installations as blue used to be

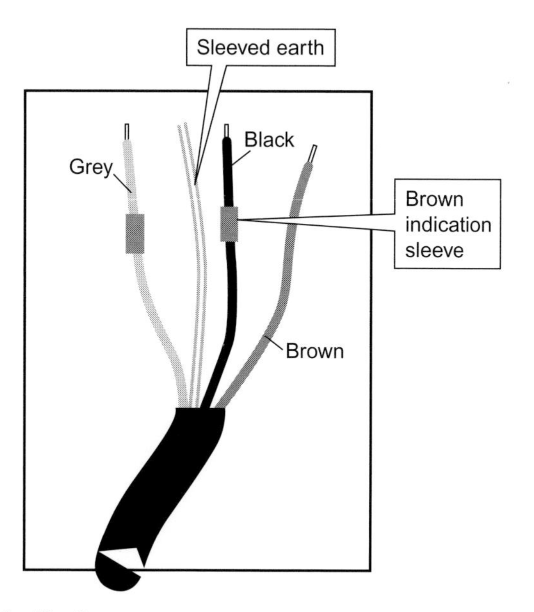

FIGURE 5.2 **Conductor identification**

the colour of the blue phase. Line conductors in a single-phase supply are now brown but they used to be red.

Our old three-phase systems used to be identified as red, yellow and blue, or L1, L2 and L3, but now we must use brown, black and grey, or L1, L2 and L3. Where black or grey conductors are used as line conductors in single-phase installations, they must be identified by using a brown sleeve at the termination which only needs to cover a small visible section of the cable, or if they are being used as a neutral then they must have a blue sleeve at the termination on a small visible section of the cable (Figure 5.2).

A protective conductor must be identified and sleeved with green and yellow for its complete length where it is terminated at an accessory, or if it is a single conductor it must be green and yellow for its entire length. It is also a requirement that the colouring of the sleeving is green and yellow; one of the colours must be a minimum of 30% and a maximum of 70%. Once again if we use materials to a British Standard, we will have no problems.

The single colour of green is no longer acceptable for identification and if it is found in an installation during a periodic inspection it needs to be listed as an observation at category 4.

A PEN conductor can be green and yellow and identified at one end with blue sleeving, or it can be blue and identified with a green and yellow sleeve.

Diagrams or charts need to be provided giving information on the installation. Completion of the correct certification from appendix 6 of BS 7671 will satisfy this requirement.

Where a voltage exceeding 230V is present where it is not expected, a warning notice must be provided. This could be where socket outlets on different phases are close together or where two phases are present within a switch. This type of notice is not a requirement where a voltage greater than 230 V is expected and would include items of equipment such as bus bar chambers or distribution boards.

Isolators need to be identifiable where it is not obvious which piece of equipment they are intended to isolate or where a piece of equipment cannot be isolated from a single position.

Notices are also required for the following:

- Periodic inspection dates
- RCD quarterly test
- Earthing and bonding conductors with labelling to BS 951. Where there is a number of earth clamps in close proximity to one another (*places such as airing cupboards*) not every point of connection needs to be identified as it will be obvious that they are earth clamps
- Non-conducting locations
- Harmonisation of colours
- Dual supply
- High protective conductor currents
- Electromagnetic capabilities.

515 Prevention of mutual detrimental influence

Whenever electrical equipment is installed, as well as ensuring that it is compatible with the rest of the installation, we must also take care that it does not have a harmful effect on any non-electrical services and that none of these services has a harmful effect on our installation. This could be as simple as keeping our cables away from central heating pipes, or making sure that parts of our installation are at a reasonable distance away from services which may require maintenance in the future. (*No bonding conductors tied to parts of the gas or water installation.*)

How often have you seen cables pushed between or next to central heating pipes laid under floors? Quite often I expect, but it is not very good for the cables.

Segregation between our installation and data cables, phone lines, etc., is also part of this requirement.

CHAPTER 52

Selection and erection of wiring systems

The selection of the wiring system and the way in which it is installed are dealt with in this section of the Wiring Regulations. One of the main considerations is that any electrical equipment is installed in the way in which it is intended; also that any manufacturer's instructions are understood before the installation of any equipment and materials is undertaken.

521 Types of wiring system

This section of the regulations expands on the requirements of the fundamental principles as set out in chapter 13. Table 4A1 sets out the requirements of the regulations with regard to installation methods of conductors and cables.

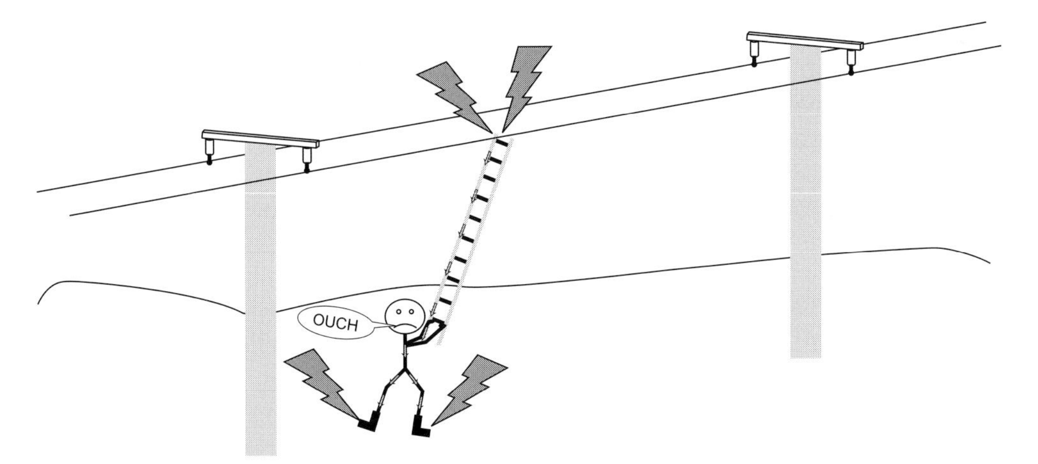

FIGURE 5.3 **Out of reach**

- **Bare conductors**: These can only be installed using insulators, and if they are bare they will need to be placed out of reach (Figure 5.3).
- **Non-sheathed cables**: Can be installed in conduit, trunking or ducting, or on insulators. A non-sheathed protective conductor can be installed directly on to a cable tray or clipped direct.
- **Sheathed multicore and single-core cables**: These can be installed using any method that is suitable for the installation.

Table 4A2 of appendix 4 provides a list of 77 installation methods which are given a reference letter from A to G. You can see that although there are a lot of installation methods, there are only seven reference methods, apart from the four given to flat twin and earth cables installed in thermal insulation. These reference methods are numbered from 100 to 103. All of these reference methods are used to determine the current carrying capacity of a cable for the way in which it is to be installed.

521.5 Electromechanical effects in a.c. circuits

The use of single-core SWA cable is not permitted in a.c. circuits.

Where singles are used, the conductors of each circuit must be contained in the same ferrous metal enclosure; this applies to the earthing conductor of the circuit. Conductors of the same circuit must also enter an ferrous metal enclosure through the same opening. It may look neat to have the same colour conductors through the same opening, but there is a very high risk that the magnetic effects caused by current flow will create a lot of heat and damage the cables.

Multiple circuits are permitted in the same enclosure or even within the same cable if the following rules are applied.

- Every cable or conductor is insulated to the highest voltage present. Where conductors or cables are not insulated to the highest voltage present but are insulated to the voltage required for their own system voltage, they can be installed in the same enclosure. The enclosure must have separate compartments for each voltage present. This would only be possible where trunking was being used. If a conduit system were to be used then separate conduits would be required.

- Where cables of Band I and II circuits are to be installed on a cable tray, separation must be provided by using a partition.
- In multicore cables, Band I and II circuits must be separated by an earthed metal screen which must have a current carrying capacity of the largest core of the Band II circuit.

Circuit arrangements

The line and neutral of each circuit must be separate from other circuits. It is not permitted to 'borrow' a neutral from one circuit for use in another. This bad practice can often be found in two-way lighting circuits and central heating control systems and can be very dangerous as isolation of one circuit is not possible.

Flexible cables or cords

In reality, there is little to be gained by using flexible cords or cables for fixed wiring, but it is possible. Flexible cables are really intended to be used as a final connection from the fixed wiring of an installation to an item of electrical equipment, such as an immersion heater or electric radiator. BS 7671 does allow the use of flexible cords or cables provided the requirements of the regulations are met. It would be perfectly acceptable to use flexible cables and cords in most instances, provided all of the calculations which are required for the other types of cables are used, such as $R_1 + R_2$ values and disconnection times. This will limit the length of the circuit.

Compliance with the regulations requires that consideration is given to the temperature rating of cable insulation and the temperature to which the conductor could rise in the event of a fault.

Installation of cables

All cables must be installed in a suitable manner to suit their type of construction. Single (non-sheathed) cables must be installed in trunking or conduit. The enclosure must have an IP rating of at least IP33 and the lids/covers must not be able to be removed without the use of a tool or other deliberate action.

522 Selection and erection of wiring systems in relation to external influences

This section provides a complete list of external influences that may affect an installation. A concise list of categories can be found in appendix 5 of BS 7671.

Most of the external influences listed here are self-explanatory, but where there is anything unusual to consider an explanation will be given.

- **Ambient temperature** (AA): High and low temperatures must be taken into account.

- **External heat sources**: A wiring system must be protected from the effects of external heat sources. The correct temperature rating of cable must be used, for example, when connecting an immersion heater.
- **Presence of water** (AD) **or high humidity** (AB): IP ratings must be correct and protection from ingress of water and corrosion must be provided where required.
- **Presence of solid foreign bodies** (AE): This is protection against dust and ingress of other objects. IP ratings are used to describe the level of protection required. Consideration must also be given to any area where a build-up of dust may prevent heat dissipation. It is quite permissible to provide a system where dust removal is possible if required.
- **Presence of corrosive or polluting substances** (AF): Corrosion is caused not only by dampness, but also by chemicals and acid in animal urine. Suitable protection must be provided where any of these could be a problem. Protection can be provided by covers, paint, grease or even applying a layer of tape. Electrolytic action will also cause damage where dissimilar metals are touching, for instance an unsheathed copper mineral-insulated cable clipped to an aluminium surface.
- **Impact** (AG): Protection from impact can be provided in many ways and the level of the severity of any impact must be assessed. This section also deals with cable penetration. The basic rules are that if a cable is placed under a floor or above a ceiling it must be at least 50 mm from the surface or have an earthed metal covering or be mechanically protected from penetration by nails and screws, etc. There are various methods of protection for a cable concealed in a wall or partition. It must be at least 50 mm deep or mechanically protected against penetration by nails and screws, etc. It can also be enclosed in earthed metal such as steel conduit. Where any of these methods are used, the cables can be concealed anywhere within a wall or partition.

  We can also install our cables within zones. Provided that the cables are installed in the required zones, protection can be provided by the installation of an RCD with a trip rating of 30 mA. The zones are 150 mm from the ceiling or internal corner of a wall and vertically and horizontally from the accessory. The zone also extends through any wall which is 100 mm or less thick (Figure 5.4).

  Cables clipped direct have no requirements other than to ensure that the mechanical protection offered is suitable for the zone.

  Where a cable is installed in a wall or partition which has metal studs, unless it is protected by an earthed metal enclosure or mechanically protected against damage, 30 mA protection must be provided.
- **Vibration** (AH): Vibration can have many effects. How many times have you gone to use a socket outlet or a switch which has been in your van for some time only to discover that the terminal screws are missing?
- **Mechanical stresses** (AJ): Whenever a cable is installed, we have to take care that it is not damaged during the installation process. For this reason, we have to ensure that the wiring system such as conduit or trunking is completed before

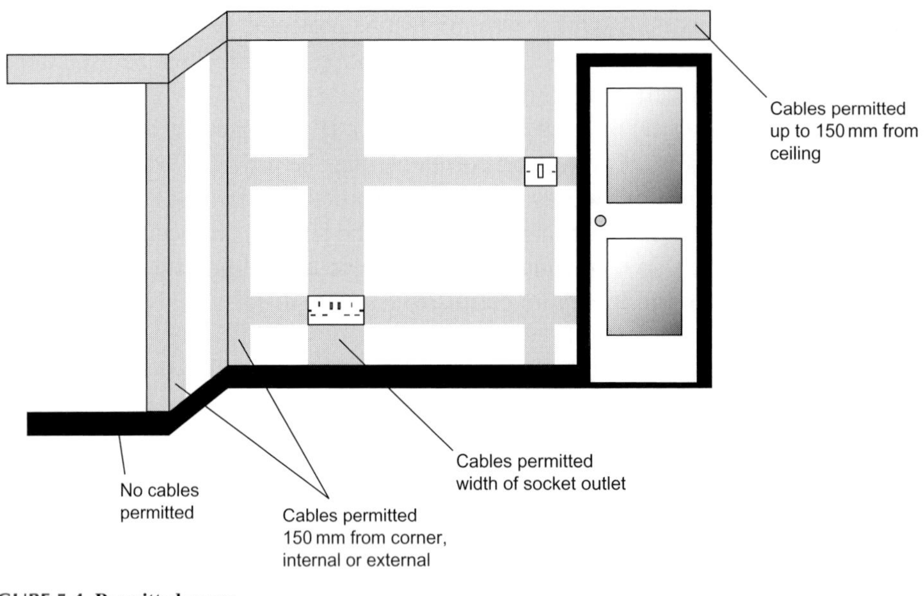

FIGURE 5.4 **Permitted zones**

any cables are introduced into it. The system must also have adequate drawing-in points and be free from sharp edges.

Cables need to be supported suitably. The correct support spacings can be found in the on-site guide or guidance note 1.

Where cables are not supported, they can suffer damage caused simply by their own weight. Cables installed vertically, perhaps in a trunking system, must be protected from damage by the use of pin racks. Lack of support is one reason why cables should not be installed in cavity walls. Care should also be taken to ensure that damage does not occur where a cable changes direction from horizontal to vertical within a trunking system as bending radii are also very important.

- **Presence of flora and/or mould growth** (AK): Plants and mould can cause all sorts of problems. For instance, if a plant were to grow around a PVC conduit system, the weight of the plant could damage the conduit or in extreme cases pull it completely away from its fixings. It could also cause corrosion by keeping a system damp. Mould could grow on parts of the installation which again would create damp areas and cause corrosion.
- **Presence of fauna** (AL): We have all seen cables within an installation which have been chewed by mice and rats, even squirrels. Not a lot can be done to prevent this kind of damage from happening.

  Where we know that wildlife or insects are likely to have a harmful influence on any parts of an installation, we need to ensure that we protect against damage. Protection can be mechanical or the use of appropriate IP codes, particularly where insects are present. The location should also be chosen carefully where the installation permits it.

- **Solar radiation** (AN) **and ultraviolet radiation**: Certain types of cable will be damaged by continued exposure to sunlight, as will some types of PVC conduit. Where cables are to be fixed in direct sunlight, I have always found FP 200 to be suitable. The temperature rise of the conductors needs to be taken into account when selecting the conductor size.
- **Seismic effects** (AP): This refers to earth movement and volcanoes and is unlikely to affect us in the UK.
- **Wind** (AS): This can be treated the same as vibration and mechanical stress. We need to ensure that any equipment which may be affected by high winds is fixed securely.
- **Nature of processed and stored materials** (BE): Where materials which could burn easily are being worked on or stored precautions must be taken to prevent the spread of fire.
- **Building design** (CB): Where a building is likely to move due to the structure of the building, allowance must be made for this movement. This could be where different rates of expansion could occur, or even the wiring inside a tent or marquee.

523 Current carrying capacity of cables

This section of the regulations provides guidance on the selection of the correct size of conductor with regard to its current rating.

Heat is the enemy of any electrician as it will cause damage to cables and could result in a fire. Most of the guidance here relates to ensuring that cables do not overheat. The maximum permissible operating temperatures of cables are dependent mainly on the type of insulation material used in the cable construction and are:

Thermoplastic, maximum temperature of the conductor = 70°C
Thermosetting, maximum temperature of the conductor = 90°C
Mineral, thermoplastic covered or bare and exposed to touch = 70°C
Mineral, bare and not exposed to touch and not in contact with combustible material = 105°C.

Care must be taken to ensure that any conductor which is operating above 70°C is terminated into equipment that can withstand the higher temperature, as the British Standard only requires that our equipment is suitable for 70°C.

Ambient temperature must always be taken into account. The current carrying capacities given in the tables in appendix 4 are calculated for a cable operating in an ambient temperature of 30°C. For example, if we look in table 4D1A for a 4-mm^2, we can see that a copper conductor clipped direct can carry a current of 37 A at 30°C. If we were to pass a current of 37 A through the conductor it would rise to 70°C, which is fine. If the ambient temperature were to increase to say 40°C, then the temperature of the conductor when we passed 37 A through it would rise to 80°C. A side effect of this would be an increase in voltage drop.

Groups containing more than one circuit

Temperature is again the factor which dictates the size of the conductor in relation to the number of circuits installed. Where cables are grouped together, the dissipation of heat is affected. Tables 4C1 to 4C5 give rating factors to be used for cables grouped together. If it is known that a cable will not be carrying above 30% of its grouped current rating, it can be excluded from the grouping calculation.

Example

A single-phase circuit with 2.5-mm^2 cable is installed in a conduit fixed directly to a wall, the conduit contains two other circuits and the ambient temperature is 30°C.

From table 4D1A, reference method B we can see that the conductors can carry 24A (I_t). To calculate the current that this cable can carry under its installation conditions, we need to look in table 4C1 for its rating factor. The rating factor for three circuits in conduit is 0.7.

Our calculation is $I_z \leq I_t \times$ rating factor.

$$24 \times 0.7 = 16.8\text{A } (I_z)$$

If our cable is not carrying more than 30% of this value, 16.8 × 30% = 5.04, it need not be included in the grouping calculation.

This sort of calculation is often not much use unless larger cables are being used to compensate for voltage drop.

Cables in thermal insulation

Table 52.2 provides derating values for cables which are surrounded by thermal insulation. Provided we use the correct tables in appendix 4 when we are selecting the current rating for flat twin and earth cables we do not have to refer to table 52.2.

For other types of installation such as conduit, we would have to apply the rating factor from table 52.2. It shows that any cable which is completely surrounded by thermal insulation will have its current rating reduced by 50%.

Example

From table 4D5, a 2.5-mm^2 conductor installed to reference method C is able to carry 20A. If it was completely surrounded by thermal insulation, it would be installed as in reference method 103 and the current carrying capacity would now be 10A.

Variations of installation conditions along a route

Occasionally, the circuit may be affected by more than one condition that would affect the dissipation of heat. Where more than one installation condition applies at the same time all of the appropriate rating factors from appendix 4 must be used; however, if they apply to the circuit at different sections of the cable route then only the worst case applies (Figure 5.5).

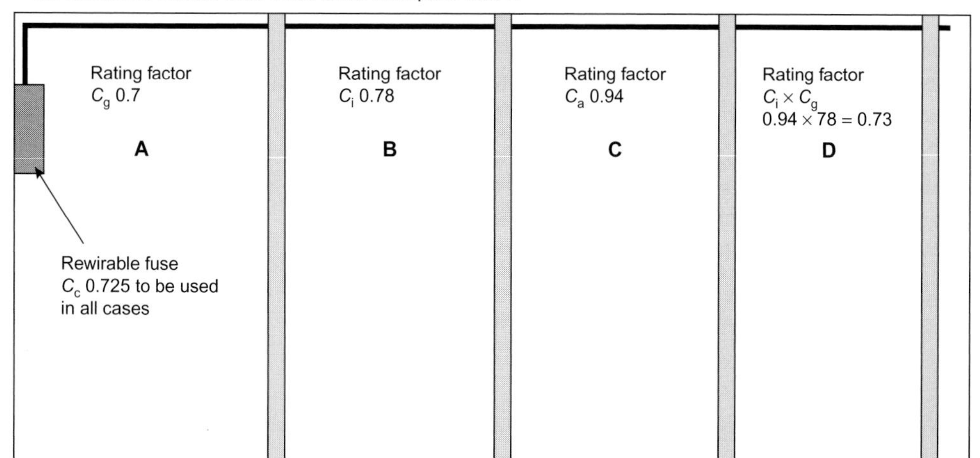

FIGURE 5.5 Rating factors affecting cables

TABLE 5.1 Voltage drop

230V supply	400V supply
Lighting 3% = 6.9V	Lighting 3% = 12V
All other circuits 5% = 11.5V	All other circuits 5% = 20V

Cross-sectional areas of conductors

The minimum size of conductor that we can use in most of our installations is 1 mm^2 if copper and 16 mm^2 if aluminium. Flexes used for appliances should be as recommended by the manufacturer.

Where aluminium conductors are used, they must be tested and approved for use with aluminium.

Voltage drop in consumers' installations

Appendix 12 provides us with information regarding voltage drop.

We are permitted a voltage drop of 3% for lighting circuits and 5% for all other circuits in installations which are connected to the public supply. See Table 5.1.

These voltage drops are in addition to those permitted by the supply, which can be found at the end of appendix 2. For a 230-V supply the permitted tolerance is +10%/ −6%, which is 216.2/253 V. If the installation is supplied by a private low-voltage generator, the total voltage drop from the generator to the load can be lighting 6% and all other circuits 8%. The total voltage drop for each final circuit must not exceed 3%/5% depending on the type of circuit.

Where a voltage drop of short duration which is greater than permitted is caused in a circuit due to startup or inrush currents, it may be ignored provided it is within the limits acceptable to the equipment manufacturer.

526 Electrical connections

All electrical connections have to be made within a suitable enclosure. Where connector blocks are used they must be enclosed; wrapping tape around them would not satisfy the regulations.

All terminals must be of a suitable size to receive the number and CSA of the conductors that require termination. They must also take into account the shape, material and type of insulation of the conductors.

Occasionally, you will come across a terminal which is marked:

- R is for rigid conductors only
- F is for flexible conductors only
- S or Sol is for solid conductors only

Where a terminal is unmarked, it is accepted that it is suitable for all types of conductor.

Where vibration or changes in temperature may occur then security/locking of the terminations must also be considered as these conditions could result in the loosening of terminations.

As a general rule, all connections must be accessible for inspection, testing and maintenance. It is a good idea to leave a plan of where any junction boxes can be found within an installation. It is perfectly acceptable to put a junction box under a floor or in a loft space provided it can be found and accessed in the future. The exception to this would be where a joint of a permanent nature is made, such as:

- Compound-filled joints or joints designed to be buried
- Connection between a cold tail and a heating element for floor or ceiling heating
- Compression, welded, soldered or brazed joints.

Of course, before any of these connections are covered they should be very carefully examined and tested. It is also a good idea to leave a plan of where the joints are.

Where the connection of fine or very fine wire is required, it is important to ensure that the correct types of terminals are used or the use of bootlace ferrules is considered (Figure 5.6).

Loose connections can occur due to changes in temperature caused by current flow.

FIGURE 5.6 Ferrule

Where screw connections are used, the ends of the fine wire conductors should not be soldered. It is very easy to screw through the conductor without being aware that this has happened.

527 Selection and erection of wiring systems to minimise the spread of fire

Precautions within a fire segregated compartment

At first glance, this could be mistaken to mean an electrical compartment, when in fact it is a building compartment. The objective of this set of regulations is to prevent the spread of fire in areas where there is a particular risk of fire.

Where cables are installed in enclosures such as conduit and trunking no particular requirements are needed provided the enclosures meet the required British Standard. Where cables are to be clipped direct they must comply with the requirements of BS EN 60332-1-2 or BS EN 50266. Most thermoplastic cables such as 6242Y or 6491X will not meet these requirements, whereas 6242B and 6491B will be suitable. If in doubt it is better to seek the advice of the cable manufacturer by a phone call, or there is a lot of valuable information on the Internet. All other equipment used in these areas will also need to comply with the required flame propagating standards.

527.2 Sealing of wiring system penetrations

Any wiring system that passes through a part of the building which is a fire barrier must be sealed to the same degree of protection as is offered by the fire barrier. Where the wiring is in an enclosure which has a large internal area of greater than 710 mm^2 (27 mm × 27 mm), the system must be internally sealed.

Where a wiring system passes through a fire barrier it must be sealed by a material that offers the same degree of fire protection as the barrier itself. The material must be compatible with the wiring system and the barrier and it must also offer the same degree of protection from penetration of water. Due to temperature changes, the wiring system may expand or contract. The sealing arrangement must also allow for thermal movement.

The wiring system on either side of the sealed fire barrier must be suitably supported. This is to ensure that in the event of a fire on one side of the barrier, if the cable were to collapse due to the melting of the support system, the weight of the cable would not damage the fire seal.

528 Proximity of wiring systems to other services

If it is required that Band I and II are contained in the same wiring system one of the following measures must be taken.

- All conductors must be insulated to the highest value present.
- A separate compartment must be provided for each voltage present.
- If installed on a cable tray, physical separation must be provided by a partition.

- The circuits may be in the same multicore cable but they must be separated by an earthed metal screen with a current carrying capacity of the highest conductor present.

528.2 Proximity of communication cables

Where parts of our installation are buried and cross telecommunication cables, we must ensure that the cables are separated by 100 mm. In an instance where this distance cannot be met we can separate the cables by a fire-retardant partition (bricks, cable tiles, etc.).

Mechanical protection between the cables must be provided at crossings to prevent damage to cables due to heavy traffic.

528.3 Proximity to non-electrical services

Compliance with this regulation requires that we ensure that wherever we install any part of a wiring system, we take precautions to ensure that it causes no harm to any non-electrical services, such as heat, vibration or electromagnetic interference. We must also ensure that the installation is only installed in areas where it is permitted; for example, lift shafts would be an area which must be avoided. Consideration must also be given to any non-electrical services which could cause damage to the installation. This could be due to maintenance being carried out on a water system or by condensation. Protection could be provided by any means which are suitable for the situation.

529 Selection and erection of wiring systems in relation to maintainability

Wherever we install any electrical equipment, there is a good chance that it will need to be accessed and isolated for maintenance.

Where isolation has to take place, the installation must be constructed so as to ensure that access is possible, and that after any isolation the installation can be reinstated without damage.

CHAPTER 53

Protection, isolation, switching, control and monitoring

Four types of isolation and switching are recognised by the Wiring Regulations. These are:

- Functional switching
- Isolation
- Switching off for mechanical maintenance
- Emergency switching.

Functional switching

Functional switching is an operation intended to switch on and off the supply of electrical energy to all or part of an installation for normal operating purposes.

This also includes devices for variation of the supply, such as dimmer switches. In some instances, they can be used for mechanical maintenance.

This type of switching is widely used to control individual sections of an installation or to control a number of items of equipment. A switch can be used to control all live conductors (line and neutral); this needs to be a double pole device. A single pole device can be used but must only be connected in the line conductor.

A functional switch must be capable of breaking the highest current which is going to pass through it under normal conditions without presenting a dangerous situation. A functional switch must be suitable for the external influences which exist in the area where it is installed. It must also be identifiable and in some instances labelled, particularly where its function is not obvious.

Equipment such as circuit breakers, RCDs, plug and socket outlets up to 32A may be used as functional switches. It is possible to use some isolating switches as functional switches, but care must be taken to ensure that they are intended to switch while on load. Items such as off-load isolators and fuses should never be used for functional switching.

Isolation

Isolation is intended to cut off the supply for safety reasons; this could be for the complete installation or sections of the installation as required. Where the isolator is not visible or is remote from the equipment it is being used to isolate, it must be manufactured in such a way that it can be secured in the off position by a locking device.

There is a difference between an isolator and an isolating switch.

An isolator is intended to isolate a piece or all of the installation when it is off load; it is not intended to be used as a functional switch. Where this type of device is installed precautions must be taken to ensure that only electrically skilled persons have access to it. It must also be identified as being unsuitable for switching on load. This is because if it is used to switch a load it may cause damage. It must also be capable of being secured to prevent unauthorised operation.

An isolating switch, which is often called a switch disconnector, can be used for switching while on load as it is designed for making, carrying and breaking currents under normal load conditions. This type of device can be used for functional switching where required (Figure 5.7).

Part 2 of BS 7671 gives the following definition:

A mechanical switching device capable of making, carrying and breaking currents under normal circuit conditions which may include specified operating overload conditions and also carrying for a specified time currents under specified abnormal circuit conditions, such as those of short circuit.

Switch symbol

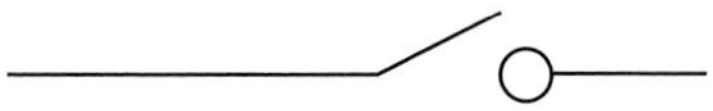

The symbol ('O') indicates the suitability of the device for making and breaking of a current carrying circuit, in other words a functional switching device (load-break or on-load switch). However, it should be noted that this symbol does not indicate suitability for the purpose of isolation.

FIGURE 5.7 Switch

Certain circuit breakers may be used for isolation. The symbol for this type of circuit breaker is shown in Figure 5.8.

Isolation can be achieved by using various types of switching device:

- Disconnector symbol (Figure 5.9)
- Switch fuse disconnector (Figure 5.10)
- Fuse switch disconnector (Figure 5.11)
- Fuse combination switch (Figure 5.12).

Each installation must have provision for disconnection from the supply. Where possible, it is better to have one single device which will isolate the entire installation; however, some installations will have an off-peak supply. This type of arrangement will require the use of separate switches which will have to be labelled.

A TN single-phase system which is under the control of skilled or instructed persons may be isolated by the use of a single pole switch. We often use a single pole protective device for isolation when we are working on a circuit. Of course, this is acceptable as we are deemed to be skilled persons. Where the installation is to be under the control of ordinary persons, or is a TT or IT system, isolation must be provided by a double pole device; domestic installations are an example of this.

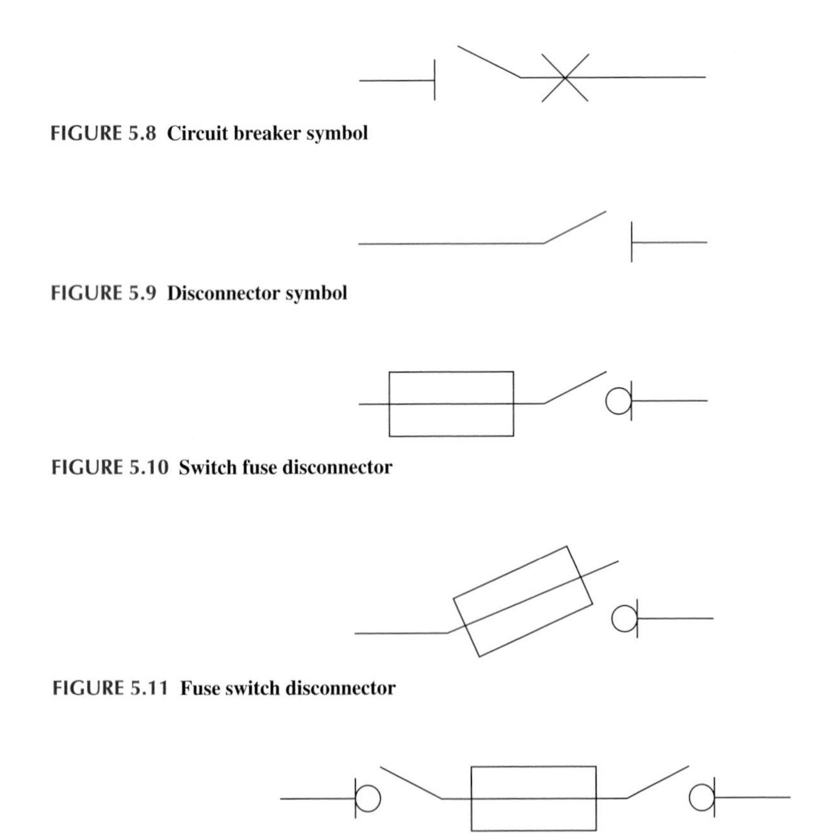

FIGURE 5.8 Circuit breaker symbol

FIGURE 5.9 Disconnector symbol

FIGURE 5.10 Switch fuse disconnector

FIGURE 5.11 Fuse switch disconnector

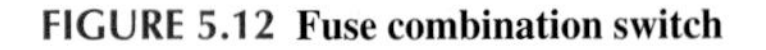

FIGURE 5.12 Fuse combination switch

TABLE 5.2 Requirements to switch the neutral conductor

System	Single phase	Three phase
Main switch		
TN-C-S	Yes	No
TN-S	Yes	No
TT	Yes	Yes
Downstream isolation		
TN-C-S	No	No
TN-S	No	No
TT	Yes	Yes

In three-phase systems, there are no allowances for ordinary persons, and isolation of the supply on TN systems must involve the disconnection of all line conductors but not the neutral, although it must be possible to disconnect the neutral by the use of a link which can only be removed with a tool.

Where the installation is supplied by a TT or an IT system, isolation of the supply must be on all live conductors (Table 5.2).

Neutral isolation would also be required for the following:

- Areas where particular risks of fire exist
- Uninterrupted power supplies
- Self-contained generation systems
- Areas which have potentially explosive atmospheres.

The neutral should always be isolated where it is necessary to prevent danger as it may have a potential which is above zero. This potential could in some situations be enough to give a small electric shock. A small shock may not be dangerous in most cases and would just be uncomfortable for the recipient, not unlike touching the probes of an insulation resistance test instrument. Problems could occur when working on a ladder or a pair of steps, as the small shock could be enough to cause a loss of balance and a fall.

There is no regulation which prevents the isolation of any neutral providing it is not a PEN conductor.

Switching off for mechanical maintenance

A device which is to be used for switching off for maintenance must have the off position clearly indicated. In most cases, it should be able to be secured in the off position and unless there is no possibility that it could be confused with other devices, it must be clearly identified.

Common sense has to be used because something as simple as a light switch could be used when changing a lamp or cleaning a light fitting; this would be classed as mechanical maintenance. In this type of situation, no other precautions would be necessary as live parts would not be exposed.

Devices for mechanical maintenance must also be capable of on-load switching due to the fact that they may be used by non-electrically skilled persons.

There appears to be some confusion in BS 7671 with regard to the use of plugs and sockets for mechanical maintenance. Table 53.2 clearly shows that a plug and socket outlet up to and including 32A may be used for isolation and functional switching; however, regulation 537.3.2.6 states that a device not exceeding 16A may be used for switching off for mechanical maintenance. I am sure that you can make your own mind up about this, but as far as I am concerned, if something is disconnected by a plug and socket it is safe to work on. Guidance note 2 does indicate that plug and socket outlets rated above 16A should be for use by skilled persons. In my experience, if an unskilled person sees a plug and needs to isolate a piece of equipment he unplugs it.

Emergency switching

Where emergency switching is required the supply must be able to be switched off by one single action (Figure 5.13).

The device or devices must be in a prominent and easily accessible position and be clearly identifiable. An emergency stop button once activated must remain latched in the off position and must only be able to be reset by a deliberate action. An emergency switch usually operates a control circuit to a contactor or motor starter, although it can be used where appropriate simply to switch the main circuit.

FIGURE 5.13 Emergency stop button

Often an emergency switch is installed into a motor control circuit. Once operated, the motor cannot be restarted until the emergency switch is reset and the motor control operated.

Plug and socket outlets must not be intentionally provided for use as emergency switching.

Functional switching

The intention of a functional switch is to provide control over parts of an electrical installation which require separate operation. A functional switch can control more than one item of equipment and must be capable of carrying the largest load that is likely to be present.

All current-using equipment must be capable of being controlled by a functional switching device. This could be something as simple as a plug and socket outlet provided it is rated at no more than 32A. Devices intended for off-load isolation only, or fuses, are not to be used as functional switches, although devices which are intended for isolation purposes can be used as functional switches if they meet the requirements for functional switching; in other words, they are intended to switch the required load and not just isolate it.

Firefighters' switch

A firefighters' switch must be installed where an electrical installation above low voltage is used outside or discharge lighting above low voltage is installed inside. The switch is normally situated on the outside of a property and should be easily accessible to the fire service (Figure 5.14).

FIGURE 5.14 Firefighters' switch

Generally, a firefighters' switch is installed adjacent to the exterior installation or near the front entrance for an internal installation. Regulations require that it is painted red, is no more than 2.75 m above the ground and switches up for off. It must also be clearly identified. A device must be provided to prevent the switch from being accidentally switched on.

Where the installation of a firefighters' switch is required, it is always advisable to contact the local fire officer and seek advice; as with anything else it is better to get it right first time.

538 Monitoring

Monitoring of the correct functioning of electrical installations is becoming more and more important, since electrical installations are becoming more and more complex. It is no longer only the light that goes out in the event of the loss of a power supply; often the losses incurred due to power failure cause untold inconvenience and have a very high financial cost.

Imagine the cost of the loss of supply to banking systems or the trauma caused when a lift breaks down; even worse would be a fire caused by faulty cables. Safety standards for electrical equipment and installations are constantly being monitored and upgraded, and residual current monitoring is beginning to be used more and more, hence the inclusion of these in BS 7671.

Insulation monitoring devices (IMDs) are for use in IT systems. IT systems are not permitted for use in the public supply system in the UK, although they may be found in some medical locations. For this reason, an explanation of the requirements of BS 7671 for IMDs will be found in the section for special installation locations.

Residual current monitors (RCMs) are based on the same principle as an RCD, where all live conductors carry the same value of current through a current transformer. If the current in the live conductors is equal, no voltage will be induced in the secondary winding of the transformer. However, with a slight leakage to earth through any live conductor, there will be a current difference in the live conductors. This current difference will be measured electronically by the RCM. When the residual current reaches a predetermined level (usually a third of the trip rating of the RCD), a signal will be sent which triggers an alarm to alert users of a possible supply failure.

RCMs are similar in operation to an RCD but without a direct switch-off function.

The characteristics of an RCM are that it will have:

- Adjustable operating values, sometimes two
- Adjustable time delay
- Alarm relays for messages
- Display LED
- LED lines or analogue indication of fault current.

An RCM can be installed at the origin of an installation or single circuits can be monitored; this is dependent on the requirements of the installation. It is even possible to monitor individual socket outlets if the situation requires it.

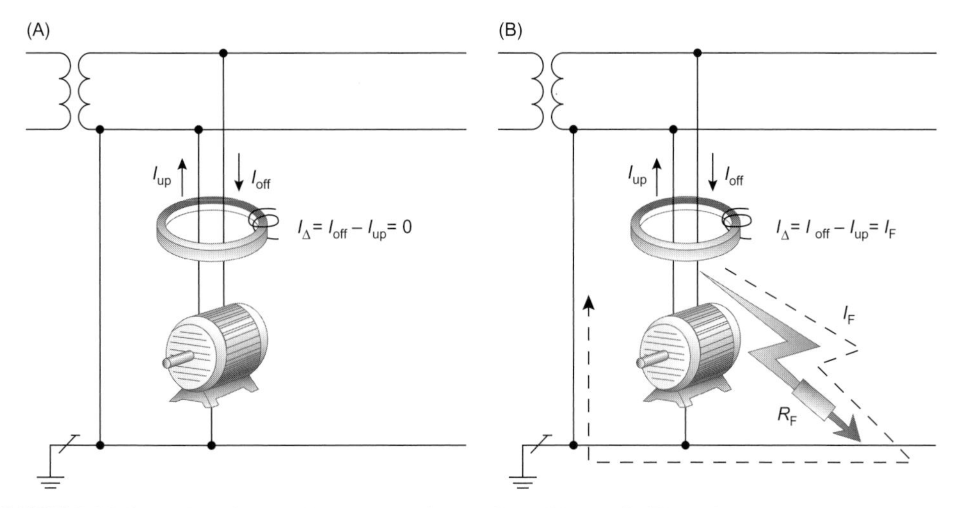

FIGURE 5.15 **Operation of residual current monitor (RCM): (A) $I_{\Delta} = 0$; (B) $I_{\Delta} \neq 0$**

Fire protection is another very good use for RCMs; there is no doubt that the first method of fire protection within an electrical installation should be careful selection. A precondition for reliability of protection and fire prevention is the correct use of protective measures for electrical installations; this includes the initial verification of the installation and then inspecting and testing as part of an ongoing maintenance programme.

The most effective protective device is one which recognises insulation faults as quickly as possible. It stands to reason that insufficient protection cannot produce a safe environment. RCDs are good for fire protection but RCMs can increase the level of protection as they can be adjusted to register even the smallest fault current (Figure 5.15).

CHAPTER 54

Earthing arrangements and protective conductors

General requirements

Within any installation, the means of earthing and bonding must satisfy the requirements of BS 7671. Where the installation is in a building which has a lightning protection system, it is important to refer to BS EN 62305.

The quality of earthing and bonding with any installation is extremely important. There is often confusion as the conductors used for either are referred to as protective conductors, both are usually connected to the main earthing terminal of the installation and they also the same colour. To get it right, we must understand the difference between earthing and bonding.

Earthing

An earthing conductor is in place to provide an easy route to earth when there is a fault between live conductors and earth. If the correct size earthing conductor is selected, it will ensure the fast operation of the device protecting the circuit and dramatically reduce, if not completely eliminate, the risk of electric shock or fire.

Earthing conductors are also used to allow the safe flow of protective conductor currents to earth in circuits such as for IT and monitoring equipment.

Those of us working in the electrical industry use all sorts of incorrect terminology which we all understand most of the time, but occasionally mistakes happen because of a misunderstanding due to the wrong terminology being used. It is common for us to hear such terms as 'earth bonding' and 'cross bonding'; as electricians, most of us know what is meant when we hear electricians' slang but it would be far easier and safer if we were to use the correct terms.

Figure 2.1 in part 2 of BS 7671 gives us the correct terminology for earthing and protective conductors (Table 5.3).

A conductor which is connected to the means of earthing is called an earthing conductor. This could be connected to the main earthing terminal and either the means of earthing supplied in a TN system or the earth electrode in a TT system. Clearly, there will only be one earthing conductor in any installation.

The earthing conductor must always be provided with a means of disconnection in an accessible position; this is to allow the measurement of the external earth loop impedance. Disconnection must only be possible by the use of a tool, in most cases a screwdriver or spanner (regulation 542.4.2).

Within any installation, it is vital for safety reasons that the correct size of earthing conductor is selected. By far the simplest method is to use table 54.7 in BS 7671.

Table 54.7 gives us an easy reference for protective conductors. The simple rules are that provided the earthing conductor is of the same material as the live conductors,

TABLE 5.3 Protective conductors

Earthing conductor	Connects the means of earthing to the main earth terminal, e.g. a supply system earth or an earth electrode
Main protective equipotential bonding conductor	Connects the main earthing terminal to any extraneous conductive parts, e.g. water installation pipes, structural steel, etc.
Supplementary protective bonding	Connects extraneous conductive parts to exposed conductive parts or other extraneous conductive parts, e.g. steel baths to electric towel rails
Circuit protective conductors	Connect the main earthing terminal to any exposed conductive parts, e.g. the metal conductive case of a washing machine or electric motor

- Any circuit of up to 16 mm^2 will require the same size earthing conductor
- A circuit of between 16 and 35 mm^2 will require a 16-mm^2 earthing conductor
- A circuit above 35 mm^2 will require an earthing conductor of a minimum of half the size of the circuit line conductor.

An example would be a standard domestic installation where the size of the meter tails is 25 mm^2. If we consult table 54.7, we can see that the size of the earthing conductor required for a circuit with a line conductor between 16 and 35 mm^2 would be 16 mm^2.

It must be remembered that the company responsible for your supply may require a minimum size earthing conductor. If in doubt ask.

It is also important to remember that any protective conductor which is not made of copper must have a minimum cross-sectional area of 10 mm^2 (regulation 543.2.3).

Table 54.7 also gives us the sizes for an earthing conductor where it is not the same material as the live conductors. This will be dealt with further on in this chapter.

542.1.8 Separate installations sharing the same supply

It is not unusual to have more than one building sharing the same supply, for example outbuildings on a commercial development or something as simple as a workshop or shed in someone's garden. These types of installation have special requirements with regard to earthing and bonding.

Regulation 542.1.8 requires that where installations sharing one supply have separate earthing arrangements, the protective conductor common to the installations has to be able to carry the maximum fault current that could flow through them.

In other words, the bonding in each building has to be selected to suit the main supply. This means that for a TT or TN-S supply the bonding conductor must be no less than half the size of the earthing conductor (Figure 5.16).

Selection of the bonding conductor for a TN-C-S supply requires the use of table 54.8, which we shall deal with later.

For the selection of an earthing conductor, we have to use table 54.7.

Example

We have a supply with 70 mm^2 line and neutral conductors. From this supply we need to supply a separate building with a 60-A supply fed by a 16-mm^2 steel wire armoured cable (we can use a single conductor for a combined earthing and protective conductor).

From table 54.7, we can see that the earthing conductor will need to be a minimum of half the size of the line conductors. This will be:

$$\frac{70}{2} = 35\text{mm}^2$$

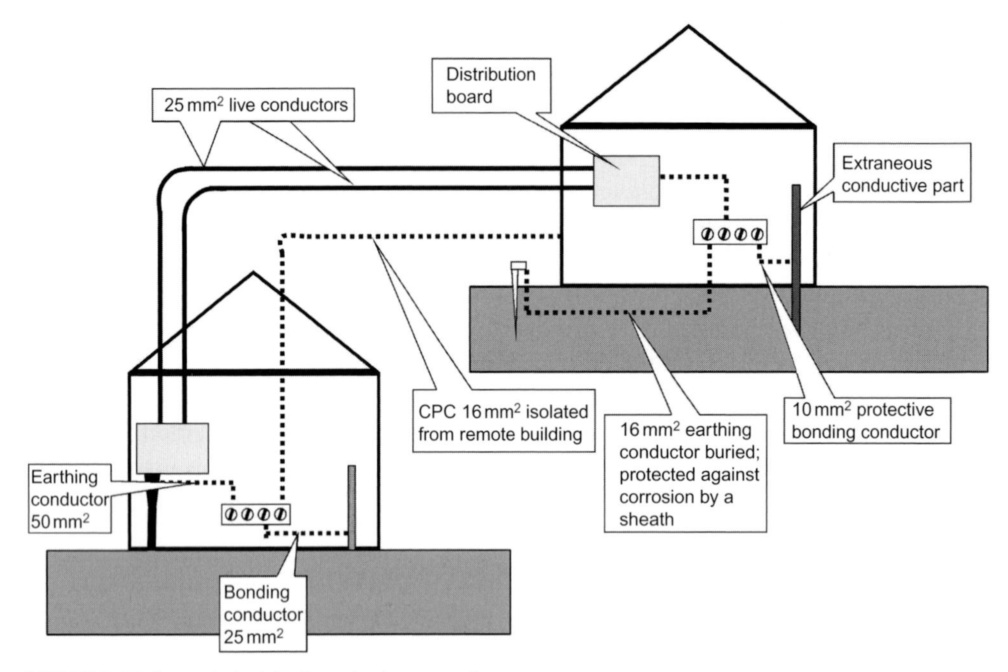

FIGURE 5.16 **Separate installations sharing a supply**

Our earthing conductor at the origin of the supply must be 35 mm^2.

To comply with regulation 544.1.1, we will have to install bonding conductors of not less than half the cross-sectional area (CSA) of the earthing conductor, $35/2 = 17.5\,\text{mm}^2$; the nearest conductor to this will be 25 mm^2 and this will have to be the size of our bonding conductor. The problem now is that if we want to use a conductor within the SWA for our combined earthing and protective conductor, we will have to use a 25-mm^2 three core when we really only require 16 mm^2 for the load.

We can use the SWA for our protective conductor provided it complies with table 54.7.

To be able to compare the results of the use of table 54.7, we will first need to consult manufacturer's data or guidance note 1 to find out the CSA of a two-core 16 mm^2 SWA.

We can see that the CSA of the armour on this cable is 46 mm^2. This would be suitable for our earthing requirements, but do not forget that we need to ensure that the armour has the equivalent resistance of a 25-mm^2 copper conductor to enable us to use it for our protective bonding. For this, we must look to see the size of the SWA which would be suitable from our manufacturer's data.

The data show that a 25-mm^2 copper conductor has a resistance of 0.725 mΩ/m and the resistance of the armour of a 16-mm^2 two-core SWA is 3.7 mΩ/m. Clearly, this resistance is too high and either a 25-mm^2 three-core will have to be used or a supplementary conductor of the full size (25 mm^2) must be installed alongside the 16-mm^2 core SWA.

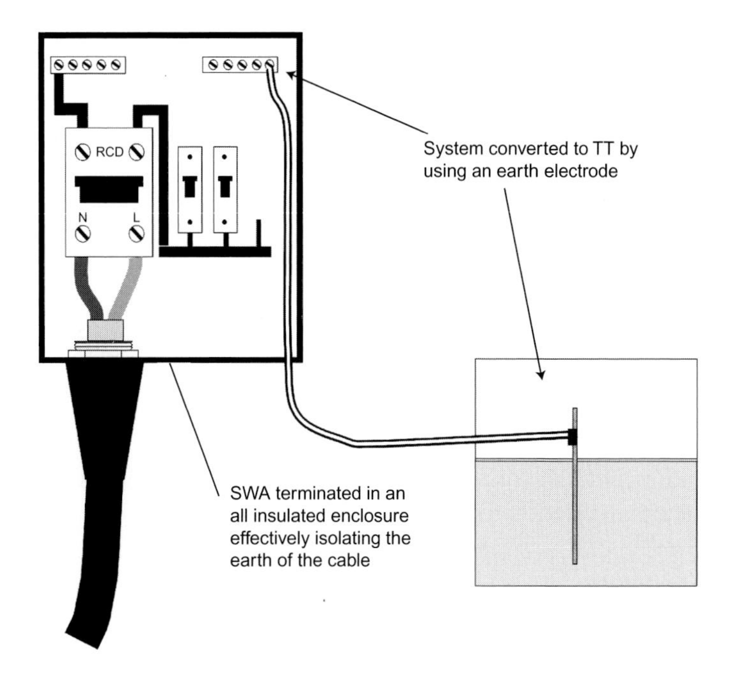

FIGURE 5.17 **System converted to TT**

From regulation 544.1.1, the maximum size protective bonding conductor required for a TN-S or TT supply need not exceed 25 mm^2.

An alternative to this would be to use a 16-mm^2 two-core SWA with the armour earthed at the supply end but isolated from the remote building. The supply for the remote building could then be treated as a TT system (Figure 5.17).

For a TN-C-S system, we have to take a different approach as we need to comply with the requirements of table 54.8.

If we use the previous example with a 70-mm^2 supply and a remote building supplied by a 16-mm^2 SWA, we can see from table 54.7 that we require a 35-mm^2 earthing conductor. We now need to look at table 54.8, which indicates that the protective bonding conductor has to be selected in relation to the size of the supply neutral (Table 5.4).

A conductor of a different metal can be used provided it has the same conductance that a copper conductor would provide.

We can see from this table that as the supply neutral is 70 mm^2, we would need a bonding conductor of 25 mm^2. From the previous example, the armour will not be large enough and either the cable will need to be a three-core 25-mm^2 cable or we must install a separate 25-mm^2 protective conductor.

We could use the option which we looked at in the previous example and isolate the earth of the supply to the remote building and convert it to a TT system. In these instances, it must be remembered that the distribution board or enclosure containing the RCD, must be constructed of insulating material as only exposed conductive parts on the load side of the RCD will be protected.

TABLE 5.4 Protective bonding in relation to supply neutral

Copper equivalent CSA of the supply neutral conductor	Minimum copper equivalent CSA of the main protective bonding conductor
35 mm^2 or less	10 mm^2
Over 35 mm^2 up to 50 mm^2	16 mm^2
Over 50 mm^2 up to 95 mm^2	25 mm^2
Over 95 mm^2 up to 150 mm^2	35 mm^2
Over 150 mm^2	50 mm^2

From regulation 544.1.1, for any TN-C-S supply we must select the protective bonding conductors in relation to the size of the supply neutral.

542.2 Earth electrodes

BS 7430 is the British Standard for earthing and contains a wealth of information for anyone who is involved in the installation of earthing systems.

Some installations require the installation of earth electrodes, which come in many shapes and forms. In fact there are not many things that cannot be used as an electrode, and it is easier to list the items which we must avoid.

- Metallic water supply pipework, unless precautions are taken against its removal. Generally, this would need to be from a privately owned water supply where the whole of the installation pipework is under the control of the owner.
- Pipework used for flammable liquids or gases.
- Steel embedded in prestressed concrete.

The type of electrode installed will depend on the type of soil and the circumstances surrounding the installation. Rods are probably the most common type used. BS 7430, the code of practice for earthing, provides us with the minimum diameter and cross-sectional area for electrodes (Table 5.5).

Where rods are installed, they must comply with the British Standard, and be properly identified and accessible for testing; they must also be contained within an inspection pit with a removable cover. Consideration must also be given to the positioning of the electrode with regard to mechanical damage and corrosion (Figure 5.18).

In some instances, it is very difficult if not impossible to use rod-type electrodes, particularly in areas which are formed of rock. In these situations, it is more practical to use plates or tapes. These types of electrodes are also very effective where there is a problem getting a low enough resistance by using rods. Plates or tapes will have a greater surface area in contact with the earth than can be obtained easily with rods.

Where plates are used, they must be of a minimum thickness of 3 mm, and the plate should be set vertically with the top being a minimum of 600 mm from the

TABLE 5.5 Minimum requirement of electrodes

Type of electrode	Cross-sectional area (mm^2)	Diameter or thickness (mm)
Copper rods or solid wires	50 mm^2	3 mm
Copper-clad or galvanised steel rods for harder ground	153 mm^2	8 mm
Copper strip	50 mm^2	3 mm
Stranded copper	50 mm^2	3 mm per strand

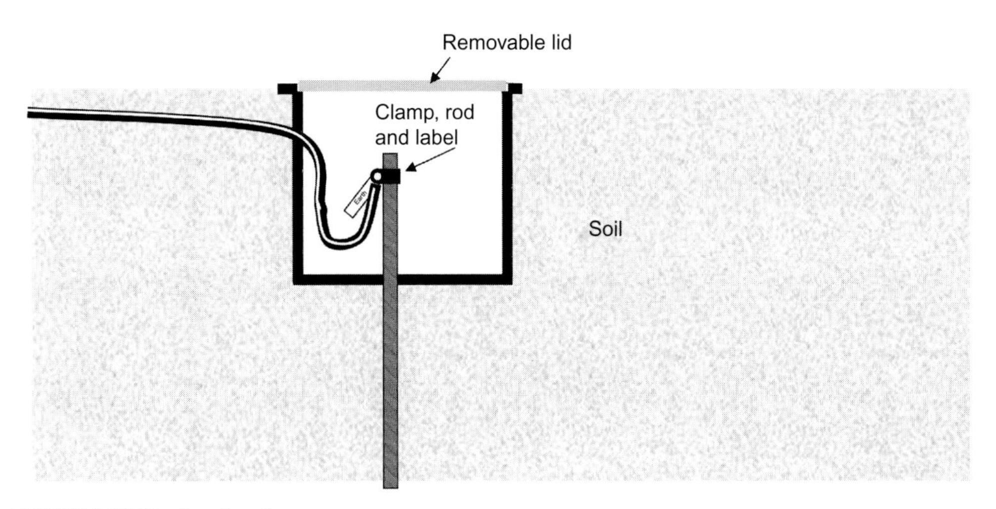

FIGURE 5.18 **Earth rod enclosure**

surface. This is to ensure that the soil which is surrounding the plate will remain damp and that in normal British winter conditions the soil around the plate does not freeze. In areas where the strata consist of rock near the surface, it is acceptable to bury the plate as deep as conditions permit (Figure 5.19).

Tape and wire electrodes should be buried to a depth of a minimum of 1 m to avoid frosts affecting the resistance of the electrode.

Wherever plates and tapes are used, it is very important that the earthing conductor is protected from damage and that the correct size of conductor is used. Table 54.1 gives us the minimum sizes which are permitted for buried earthing conductors; this table has to be used in conjunction with the requirements of protective conductors in general as set out in section 543 and where the supply is TN-C-S in section 544.

For conductors protected against corrosion by a sheath and protected against mechanical damage, the minimum size is 2.5 mm^2 copper and 10 mm^2 steel. Where they are not protected against mechanical damage they must be a minimum of

FIGURE 5.19 **Earth grid electrode**

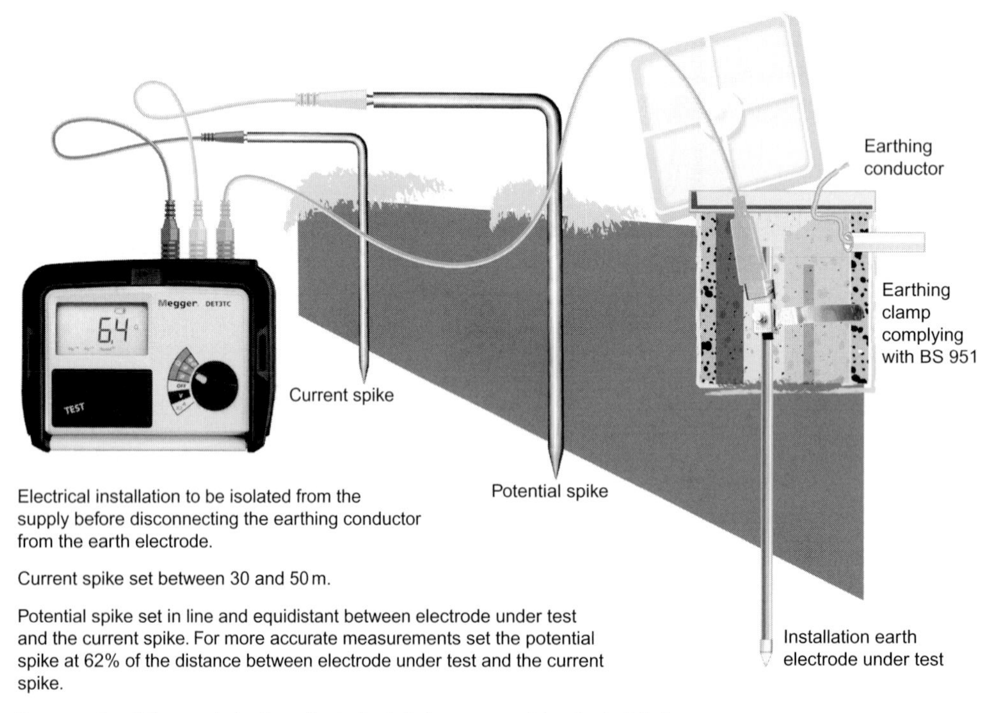

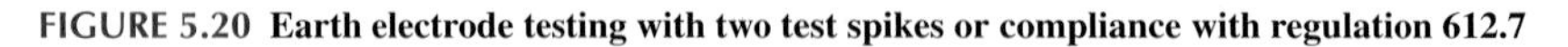

FIGURE 5.20 **Earth electrode testing with two test spikes or compliance with regulation 612.7**

16 mm^2 copper or 16 mm^2 coated steel. If they are not protected against corrosion they must be a minimum of 25 mm^2 copper or a minimum of 50 mm^2 steel.

The advantage of using plates or tapes over the use of rods is that the resistance can be reduced by adding additional electrodes in parallel and this will not have any visual impact, whereas additional rods should be spaced a distance from each other equivalent to the depth of the electrode, and this could in some instances take up quite a lot of space (Figure 5.20).

Structural steelwork will also provide an effective earth electrode and will usually give a very low resistance due to the large surface area in contact with earth. The use of this method can be quite complex as care must be taken to ensure that the earthing is not affected by corrosion. Continuity of joints in the steel must also be ensured by welding the joints which are to be below ground or placing a link on each joint above ground.

Circuit protective conductors

CPCs form part of the earth fault loop path within an installation. Its function is to carry any current which is caused due to an earth fault and it forms part of the ADS. Care must be taken when selecting the size of the conductor to ensure that it is not damaged, or causes no damage during a fault.

In most instances, these days a separate CPC is installed for each circuit whether it is installed in a metal wiring enclosure or not. This is not a requirement of the Wiring Regulations and is not necessary provided that the conduit is installed with care. A separate CPC does add an additional element of safety to the installation, but it is at the expense of using valuable space within the conduit and often results in having to install a larger size to accommodate the CPCs.

Where you choose not to install an additional CPC but are relying on the metal enclosure system to provide an earth path, care must be taken to ensure that all joints and connections are sound. One advantage of using the metal enclosure system as a CPC is that the system can be a common CPC to all circuits.

In certain cases, a separate CPC must be provided:

- Insulated wiring enclosures
- Flexible metal conduit
- Final circuits which are expected to have a protective conductor current of 10 mA or greater due to the nature of the load
- From the earth terminal of an accessory to the metal conduit or trunking system; this could be from a socket outlet box to the socket outlet.

Consideration must also be given to installing a separate CPC where there is a higher risk of corrosion than normal – in laundry rooms, food preparation areas where there are lots of stainless steel tables, etc.

Where the protective conductor is not copper, it must in all instances have a minimum CSA of 10 mm^2. Insulated sleeving must be used to cover all protective conductors up to and including 6 mm^2.

Table 54.7 must be used when selecting CPCs.

Very often the use of table 54.7 will result in the selection of a unnecessarily large conductor; this is not a problem with regard to fault protection, although it will make the installation more expensive to carry out. The use of an adiabatic equation will often result in a conductor of a smaller CSA being permissible.

The purpose of the equation is to ensure that under fault conditions the cable selected does not overheat and cause damage to its insulation or conductors.

As an example of how this works, let us carry out a simple cable selection exercise.

Example

A circuit is to be wired in 70°C thermoplastic singles installed in conduit with one other circuit. The conduit is fixed directly to a brick surface and the load is 11 kW single phase.

The circuit is 18 m long and is protected by a BS 88 fuse, it is to operate at an ambient temperature of 35°C and no other factors apply to this circuit.

The supply is 230V and has a measured Z_e of 0.4 Ω.

$$I_b = 47.82\,A$$

$$I_n = 50\,A$$

Rating factors which apply to this circuit are:

C_g from table 4C1 is 0.8
C_a from table 4B1 is 0.94

$$I_t \geq \frac{50}{0.8 \times 0.94} = 66.48$$

I_t must be equal to or greater than 66.48 A.

From table 4D1A, a conductor size can be selected; reference method B column 4 shows us that a 16-mm² conductor must be used.

Calculate voltage drop from table 4D1B.

Method A or B column 2 shows that 16 mm² has a 2.8-mV/A/m loss. We now need to calculate the total voltage drop for the circuit:

$$\frac{2.87 \times 47.82 \times 0.18}{1000} = 2.41V$$

As we are permitted 11.5V this is acceptable.

We now need to calculate the circuit earth loop impedance Z_S.

If we look in the on-site guide in table 9A, we will see that a 16-mm² copper conductor has a resistance of 1.15 mΩ/m. The object of this exercise is to select a small earthing conductor which may just give us the edge when quoting for this job. Let us try a 4-mm² earthing conductor.

Table 9A gives us a resistance of 4.61 mΩ/m for a 4-mm² conductor; therefore, the total resistance of the 16-mm² line conductor and the 4-mm² CPC will be 5.76 mΩ/m.

The resistance ($R_1 + R_2$) for the total length will be:

$$\frac{5.76 \times 18}{1000} = 0.1\Omega$$

We must also remember that the cable may be operating at 70°C; this is very unlikely but it is always best to calculate for the worst case. Table 9A gives us resistance values of conductors at 20°C. We must now make an adjustment for the rise in temperature when the cable is operational. Table 9C of the on-site guide gives us a multiplier of 1.2 for this.

$$0.1 \times 1.2 = 0.12\Omega$$

$R_1 + R_2$ at 70°C is 0.12 Ω

The total Z_s must now be calculated: $Z_s = Z_e + R_1 + R_2$

$$0.4 + 0.12 = 0.52\Omega$$

$$Z_s = 0.52\Omega$$

This value can now be used to calculate the fault current required for use in the adiabatic equation by using Ohm's law:

$$\frac{V}{R} = I$$

$$\frac{230}{0.52} = 442.3\,\text{A}$$

$$I_f = 442.3\,\text{A}$$

Next we need to look at table 3.3A in the regulations; we can use this table to calculate the disconnection time for the 50-A BS 88 fuse. Using the table, we can see that for a 0.2-s disconnection time, the fuse would require a fault current of 450 A, and our fault current is slightly less than that (442.3 A). The table shows that a fault current of 380 A will give a disconnection time of 0.4 s, and this is the time which we will use in our calculation.

$$t = 0.4\,\text{s}$$

The next step is to find the value for k; for this we must use table 54.3 of the regulations. From this we can see that the value k for a 70°C copper conductor bunched with cables is 115.

$$k = 115$$

Our calculation is as follows:

$$\frac{\sqrt{442.3^2 \times 0.4}}{115} = 2.4\,\text{mm}^2$$

The minimum CSA of the protective device is 2.4 mm^2.

This proves that our 4 mm^2 conductor will be sufficient for compliance with BS 7671, but it does not mean that we can use a 2.5-mm^2 conductor.

The adiabatic equation is used in this case as a check for compliance, not as a calculation to find the smallest conductor which could be used. If we felt that we may be able to use a conductor smaller than 4 mm^2 it would be necessary to recalculate because the R_1 and R_2 values would change, which would then alter the fault current and the disconnection time.

Earthing of accessories

There is always confusion as to whether or not accessories should be earthed to the metal back box. Clearly, if the system is steel conduit or trunking with no separate earth, an earthing tail must be fitted between the accessory and box to provide an earth for the accessory. Where the system uses a separate CPC terminated directly into the accessory, there would be no reason to provide a link between the back box and the accessory, provided the steel containment system is correctly earthed. Surface-mounted steel boxes are classed as exposed conductive parts and must always be earthed.

Flush-mounted accessories connected to a system using a separate CPC often cause the most confusion. Where the system is earthed steel conduit there is no requirement to install a link between the accessory and the box, provided the CPC is connected to the accessory. When the system is insulated conduit or twin and earth and the flush box is metal, there is no requirement to provide a link from the accessory and the box, provided the box has two fixed lugs, both of which are to be used to secure the accessory.

Where only one lug is used no earthing tail is required, provided an earthed eyelet of the accessory is used at the fixed lug position of the box. For boxes with two adjustable lugs an earthing tail must always be used.

In all instances, other than where the boxes are fixed to an earthed containment system, with a separate CPC connected to the accessory, it is always safer to provide an earthing tail and this is considered by many electricians to be good practice.

Steel wire armour used as a CPC

There is often confusion as to whether the steel wire armouring of a cable is suitable for use as a CPC. In most cases it will be fine.

Table 54.7 gives us a calculation to use which will give us the minimum CSA of the steel wire armouring which is suitable for use as a protective conductor without any further calculation other than Z_S.

Let us assume that we are going to use a 10-mm^2 two-core 70°C thermoplastic steel wire armour cable and we need to use the armour for the CPC. The calculation to find the minimum CSA of the armouring is:

$$\frac{k_1}{k_2} \times S = \text{Minimum CSA of armour}$$

From table 43.1 or 54.3, we can see that the value k for a copper conductor is 115.

It really does not matter that table 54.3 is for a protective conductor and that we are looking for value k for a line conductor; in this and most other instances, as it is a copper conductor it will have the same value k as it would have if it were to be used as a CPC.

From table 54.4, we can see that the value k for the steel sheath of the armour is 51.

Our calculation is now:

$$\frac{115}{51} \times 10 = 22.54\,\text{mm}^2$$

Provided the cable is to BS 6346 the actual CSA of the armour for a two-core 10-mm^2 cable will be 41 mm^2 and it will comply. Some of the actual armour CSAs are listed in Table 5.6. The full list can be found in guidance note 1 or from the manufacturer's specifications.

It is not uncommon for additional copper earthing conductor to be strapped to the side of an SWA cable and used as supplementary earthing. In all but very, very unusual cases this is unnecessary.

For 70°C steel wire armour cables, it is perfectly acceptable to use the armour as a CPC for all sizes up to and including two-core 95 mm^2 provided that the $R_1 + R_2$ values are suitable for use with Z_e to achieve the correct disconnection times.

From table D.9 in guidance note 1, it can be seen that all two-core cables greater than 95 mm, three-core cables above 185 mm^2 and four-core cables above 240 mm^2 would not comply with table 54.7. This does not mean that they cannot be used. It is a requirement that if the armour of these cables is to be used as a CPC the adiabatic equation is carried out; this is to ensure that the cable will comply with thermal constraints.

Let us use as an example a 120-mm^2 two-core 70°C thermoplastic SWA cable which is 53 m long buried directly in the ground. Protection is by a 200-A BS 88 general-purpose fuse. Z_e for the system is 0.1 Ω.

Reference to the table giving the CSA of armour or use of table 54.7 shows us that it cannot be used unless the adiabatic equation is carried out. This is because the required minimum CSA of the armour is 135.3 mm^2 and the actual CSA would be only 131 mm^2.

From manufacturer's data, it can be seen that the resistance of the armour on a two-core 120 mm^2 is 1.3 Ω/km. This means that it must be 1.3 m Ω/m. A 120-mm^2 copper conductor has a resistance of 0.153 Ω/km, which is 0.153 m Ω/m.

We can now use these values to calculate the expected Z_s of the circuit; this will in turn enable us to calculate the current which will flow in the event of a fault.

First, we need to calculate the resistance of the cable ($R_1 + R_2$).

$$r_1 = 0.153$$

$$r_2 = 1.3$$

Therefore, $r_1 + r_2 = 1.453$ m Ω/m and $R_1 + R_2$ for the cable will be:

$$\frac{1.453 \times 53 \times 1.2}{1000} = 0.092\,\text{mm}^2$$

The 1.2 is to correct the resistance of the conductors from 20°C to the maximum operating temperature of 70°C.

TABLE 5.6 600/1000V armoured cables: XLPE insulated, LSO H or PVC sheathed

CSA (mm²)	**300**	**240**	**185**	**150**	**120**	**95**	**70**	**50**	**35**	**25**	**16**	**10**	**6**	**4**	**2.5**	**1.5**
Cores	4	4	4	4	4	4	4	4	4	4	4	4	4	4	4	4
Diameter under armour (mm)	56.9	51.7	46	41.6	37.6	34.1	30.6	26	24.1	21.6	17.9	15.4	13.1	11.8	10.4	8.9
Overall diameter (mm)	66.8	61.2	55.3	50.6	46.4	41.7	38	32.2	30.2	27.6	24.1	21.4	19.1	16.9	15.5	14
Resistance of live conductors at 20°C (mΩ/m)	0.0601	0.0754	0.0991	0.124	0.153	0.193	0.268	0.387	0.524	0.727	1.15	1.83	3.08	4.61	7.41	12.1
Resistance of armour at 20°C (mΩ/m)	0.49	0.54	0.61	0.68	0.76	1.1	1.2	1.8	2	2.3	3.2	3.7	4.3	6.8	7.7	9.5
Minimum CSA of armour wires (mm²)	319	289	255	230	206	147	131	90	80	70	49	43	36	23	20	17
Equivalent conductance SWA/Cu (mm²)	35	35	25	25	25	16	16	10	6	6	6	4	4	2.5	2.5	1.5
Table 5.1																
CSA (mm²)	**120**	**95**	**70**	**50**	**35**	**25**	**16**	**10**	**6**	**4**	**2.5**	**1.5**				
Cores	3	3	3	3	3	3	3	3	3	3	3	3				
Diameter under armour (mm)	33.8	30.7	27.4	23.9	21.9	19.6	16.2	13.9	11.9	10.7	9.5	8.2				
Overall diameter (mm)	41.4	38.1	33.6	30	28	25.6	22.4	19.9	17	15.8	14.6	13.3				
Resistance of armour at 20°C (mΩ/m)	1.1	1.3	1.8	2	2.3	2.5	3.6	4	6.6	7.5	8.2	10.2				
Minimum CSA of armour wires (mm²)	141	128	90	78	70	62	44	39	23	21	19	16				
Equivalent conductance SWA/Cu (mm²)	16	16	10	6	6	6	4	4	2.5	2.5	1.5	1.5				
Table 5.2																
CSA (mm²)	**120**	**95**	**70**	**50**	**35**	**25**	**16**	**10**	**6**	**4**	**2.5**	**1.5**				
Cores	2	2	2	2	2	2	2	2	2	2	2	2				
Diameter under armour (mm)	27	24.7	22.3	19.3	17	15.6	15.2	13.1	11.2	10	7.7	7.7				
Overall diameter (mm)	34.8	32.3	28.5	25.3	23	20.9	21.2	18.4	16.3	15.1	12.8	12.8				
Resistance of armour at 20°C (mΩ/m)	1.3	1.4	2	2.3	2.5	3.7	3.8	6	7	7.9	8.8	10.7				
Minimum CSA of armour wires (mm²)	125	113	80	68	62	42	41	26	22	19	17	15				
Equivalent conductance SWA/Cu (mm²)	10	10	6	6	6	4	4	2.5	2.5	1.5	1.5	1.5				

We can see from the example that Z_e is 0.1 Ω; therefore, the total resistance of the circuit when operating at its full load will be:

$$Z_s = Z_e + R_1 + R_2$$

$$Z_s = 0.1 + 0.092$$

$$Z_s = 0.192\,\Omega$$

Now we can calculate the maximum fault current by using the formula:

$$\frac{U_0}{Z_s} = I \quad \frac{230}{0.192} = 11{,}979\,\text{A}$$

This is our fault current for use in the adiabatic equation.

We can also use this current to calculate the disconnection time for our 200-A BS 88 fuse.

From figure 3.3A in appendix 3 of BS 7671, we can see that the disconnection time for a 200-A BS 88 where the fault current is 1200 A is 5 s (Figure 5.21).

This will give us our time for use in the calculation.

Next from table 54.4 we can see that the value of k is 51.

Now we can complete our calculation as follows:

$$\frac{\sqrt{1263^2 \times 5}}{51} = 55.37\,\text{mm}^2$$

This is the smallest CSA of the steel wire armour which would be permissible in this instance. As we can see from our tables, the armour of the 120-mm^2 two-core cable which we are going to use has a CSA of 131 mm^2.

This cable will be suitable. If you carry out this calculation when you need to use it you will find that in nearly every case the armour is suitable for use as a protective conductor.

You may well be thinking that if this is the case why bother to do the calculation? The answer to that is simple. On the rare occasion when the armour is found to be unsuitable, the risks involved would be unacceptable, so if in doubt check it out.

In the event of the armour of a cable not being suitable due to its size we have some options; my choice in most cases would be to increase the size of the cable so that the CSA of the armour meets the required size. This also gives the option of adding additional loads, if required in the future.

The second option would be to use a cable with an additional core. Where this option is used the additional conductor must be sized to take the earth current and the resistance of the armour must not be taken into consideration.

The third option would be to install a separate green and yellow conductor to be run alongside the SWA; again this must be calculated as though it were to be the only

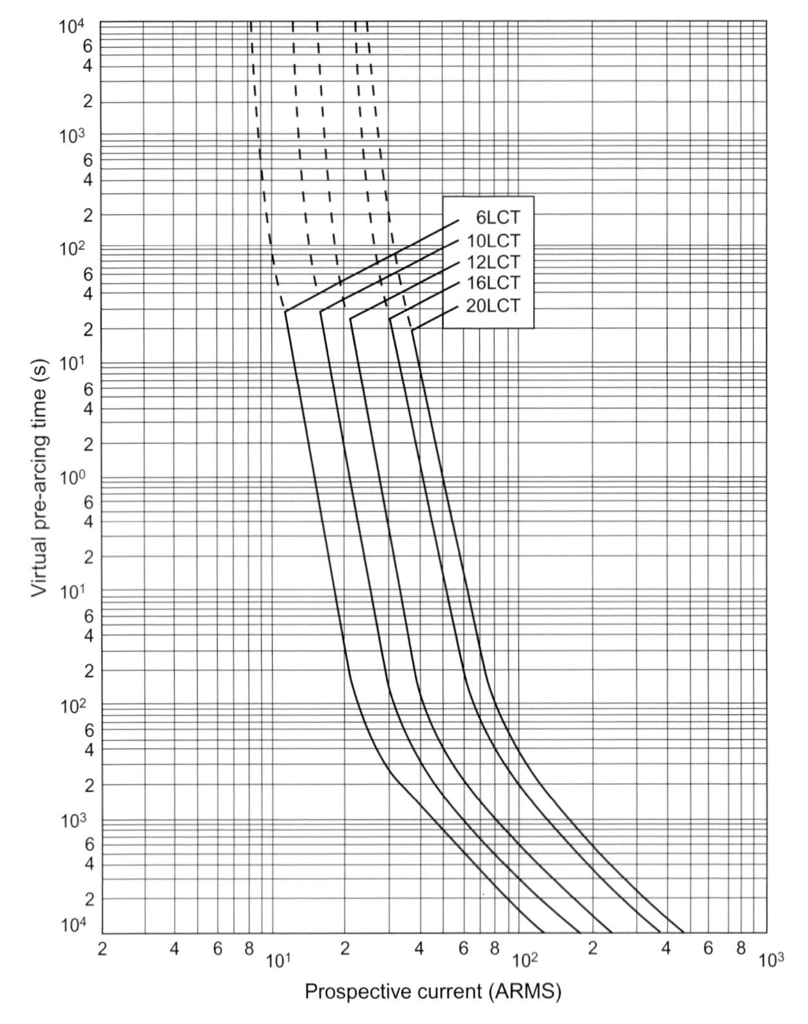

FIGURE 5.21 Time–current curves

CPC and the resistance of the armour should not be taken into consideration. This should be treated as the final option as due to the magnetic effect of the armouring, the fault current cannot be accurately predicted.

In all cases, the armour must be earthed using the correct terminations as it is classed as an exposed conductive part. Wherever possible gland tag washers (banjo washers) should be used with a copper conductor securely fixed between the washer and the earthing terminal.

543.7 Earthing requirements for the installation of equipment having high protective conductor currents

Some types of equipment require a functional earth to enable them to operate correctly; this functional earthing conductor can also act as the protective earthing conductor. In this case, it would be termed a combined protective and functional earth.

Many electricians find this section of the regulations confusing as it provides us with lots of options, and we are not always sure of what protective conductor currents are going to be present in the installation.

The simple rules are:

- For any equipment with a protective conductor current not exceeding 3.5 mA there are no special requirements.
- A single piece of equipment with a conductor current between 3.5 and 10 mA must be permanently connected to the final circuit, or connected using an industrial socket outlet to BS EN 60309-2.
- Where the equipment has a conductor current of above 10 mA, it must be either permanently connected to the supply (the final connection can be made by a flexible cable) or connected using an industrial plug and socket outlet to BS EN 60309-2.

If the equipment is to be permanently connected, the final circuit has to meet certain requirements.

- The final circuit can be a ring provided the CPC is connected into separate terminals at each socket outlet and at the main earth terminal in the distribution board. If there is more than one ring in the board it is quite acceptable to 'double up' the CPCs with those of another circuit; of course these conductors have to be identified (Figure 5.22).

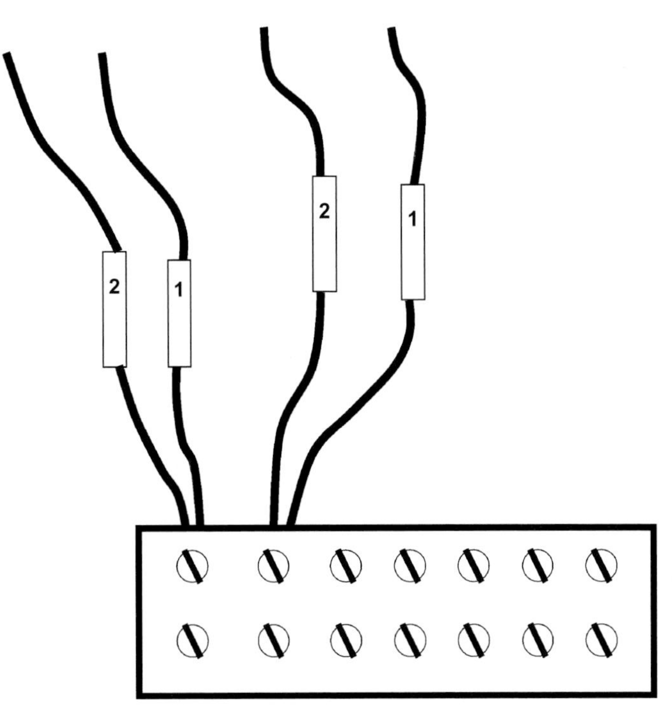

FIGURE 5.22 CPC connection for two ring circuits

- A radial circuit can be used, but the CPC would need to have a CSA of not less than 10 mm^2 copper. This large conductor is to provide a greater mechanical strength, not greater conductivity.
- Where two radials are used, it is acceptable to link the ends of the CPC at the final socket outlet of each circuit to form a loop. In this instance, the circuits must be protected by the same type and rating of protective device and the CPC must be the same size.
- A radial can be used with a single protective conductor of not less than 4 mm^2 provided the conductor is mechanically protected.

  The protective conductors can be duplicated; in other words a protective conductive can be in the form of a ring. The conductors can be individual conductors of the same material, or if it is more suited to the installation the conductors could be different types. For instance, a conductor in conduit and the conduit itself could make up the dual conductors. In some instances, it may be beneficial to use a multicore cable such as a steel wire armoured, mineral-insulated or wire-braided cable. Where this is the preferred method, the regulations require that the minimum total CSA of the cable is 10 mm^2; this includes all conductors and the armour, metallic sheath or steel braid. In reality, this will not be difficult to achieve as the armour of a two-core 2.5-mm^2 SWA cable has a CSA of 17 mm^2, and the sheath of a light-duty mineral-insulated cable has a CSA of 8.2 mm^2; add that to the 5 mm^2 of the live conductors and the minimum requirement will be achieved.
- Automatic disconnection using an earth monitoring system may be used.

TT systems

Where items of equipment with high protective conductor currents are to be installed on a TT system additional care has to be taken, not least because the protective conductor current could cause nuisance tripping.

In installations with high protective conductor currents, these currents should not exceed 25% of the RCD trip rating ($I_{\Delta n}$) for any circuit. In many cases, it is better to install one RCD per circuit rather than try and use two or three RCD for many circuits.

Where equipment having a protective conductor current exceeding 3.5 mA is installed, the regulations require that in this type of installation, twice the resistance of the earth electrode multiplied by the protective conductor current does not exceed 50 V.

$$2 \times R_A \times I_{pc} \leq 50\,V$$

Example

Let us say we have a protective conductor current of 18 mA and an electrode resistance of 80 Ω.

$$2 \times 80 \times 0.018 = 2.88\,\text{V}$$

From this calculation, you can see that in most cases this particular requirement of the wiring regulations will not present a problem, as a simple transposition of the formula will give us the maximum protective conductor current which we could allow. We know that the regulations do not allow us to use an electrode with a resistance of greater than 200 Ω (R_A), so if we use this as our worst case:

$$\frac{50}{2 \times 200} = 0.125\,\Omega$$

This shows us that even with an R_A of 200 Ω, we could still have a protective conductor current of 125 mA, which in most cases is very unlikely.

Where on the rare occasion the situation will not allow equipment to be connected directly to a TT system the equipment could be supplied by a double-wound transformer, or possibly additional electrodes could be installed to reduce the electrode resistance.

544 Protective bonding conductors

For a TT or TN-S supply, the cross-sectional area of a copper main protective bonding conductor must be not less the half of the CSA of the earthing conductor of the installation. It must be greater than 6 mm^2 but need be no larger than 25 mm^2. Other materials may be used as conductors provided they offer the same conductance as the size of copper conductor required.

For a TN-C-S supply, the protective bonding conductor must be selected according to the largest neutral of the supply.

- Up to and including 35 mm^2 requires a 10-mm^2 copper conductor
- Over 35 mm^2 up to 50 mm^2 requires a 16-mm^2 copper conductor
- Over 50 mm^2 up to 95 mm^2 requires a 25-mm^2 copper conductor
- Over 95 mm^2 up to 150 mm^2 requires a 35-mm^2 copper conductor
- Over 150 mm^2 a 50-mm^2 copper conductor may be used.

Different materials may be used provided they offer at least the same conductance.

We have to remember that the protective bonding to any services must be made as near as possible to where the service enters the building. The connection should be made within 600 mm of the service isolation point on the consumer's side. Where the meter is outside, the connection can be made within the meter box. It is important to ensure that the bonding conductor is fed into the meter box through its own dedicated entry point.

544.2 Supplementary bonding conductors

Supplementary bonding should be installed where there is a possibility of a dangerous potential difference occurring between extraneous or exposed conductive parts in the event of a fault, in particular where there is a greater risk of electric shock due to the nature of the environment; bathrooms are a good example of this.

In all cases, where a green and yellow conductor is to be used as a supplementary bonding conductor, the smallest permitted CSA is 2.5 mm^2; however, it is only permissible to use a 2.5-mm^2 conductor when it is mechanically protected, and this could be as simple as using a sheathed single cable. In most cases, it is easier to use a 4-mm^2 copper green and yellow single as this does not need any mechanical protection; care must be taken to ensure that it is installed out of harm's way and supported correctly.

Earthing clamps and labels to BS 951 must always be used and also be suitable for the environment. Where there is a possibility of corrosion the clamps should be of the blue type, or where large sizes of conductors are used then the green type of clamp may be more suitable. Red clamps are for use in dry conditions; however, blue or green can be used as these are suitable for non-corrosive as well as corrosive conditions.

Where supplementary bonding is required, various types of conductors are permitted; we are not limited to the green and yellow single cable or the CPC within a cable. It is possible to use the extraneous metal parts of the water and heating system within a building, or metal parts of the wiring system, conduit or trunking for example.

When the bonding is between extraneous and exposed conductive parts and the conductor is not copper, it must have a conductance of at least half that of the CPC connecting the exposed conductive part.

For supplementary bonding to electrical appliances, the CPC within the flexible cord would be perfectly acceptable to use as a bonding conductor provided it was connected between the earthing terminal of the accessory and the appliance.

We must not use gas pipes, oil lines or any other pipework which may hold flammable substances.

CHAPTER 55

Other equipment

551 Low-voltage generating sets

Generating sets can be stand-alone units where there is no other supply, or standby units as backup protection in the event of a supply failure. Photovoltaic (PV) systems which are also classed as generating sets are often connected in parallel with the supply.

This set of regulations covers generating sets used for mobile, temporary or permanent installations. Regardless of what the generating sets are used for the installation must comply with the requirements of BS 7671 and must be compatible with any equipment to which they are connected. Requirements for overcurrent, fault current and short circuit protection, along with current carrying capacity and voltage drop, must be considered in all installations.

For installations using a supply from generators which are not permanently fixed, possibly as a temporary supply, each circuit must be protected by an RCD with an operating ($I_{\Delta n}$) current of 30 mA maximum.

551.6 Additional requirements where the generator is used as a standby system

Where the generator is permanently fixed and is used as a backup to the normal supply it must be provided with its own means of earthing. The earthing of the supply system must not be relied upon as it may be temporarily removed under some circumstances.

Precautions must be taken to ensure that the generator cannot be used to supply the installation at the same time as the main supply. Any suitable method is permissible providing that it is foolproof.

The generating set must be compatible with the normal supply of the installation and where synchronisation is required it is simpler to use an automatic system.

A plug and socket outlet is not permissible for the connection of a standby supply and all circuits must be protected by an RCD which disconnects all live conductors.

Where a generator is being used to supply an installation which has SELV or PELV all of the requirements of regulations 414.3 and 414.4 must apply. Precautions must be taken to ensure that a loss of supply does not damage any extra-low-voltage equipment or cause a dangerous situation. This could be where supplies for safety services are required; in some cases battery backup may be required to maintain the required level of safety.

552 Rotating machines

Motor circuits only have to be capable of carrying the full load current of the motor. Starting currents do not need to be taken into account unless the motor is to be used in a situation where it is starting and stopping at such a rate that the temperature which will build up in the cable does not have time to dissipate. Under these circumstances, additional precautions need to be considered which could involve the installation of larger conductors, and possibly a higher current rating of the starter to compensate for the increase in temperature.

The general requirement for a motor circuit is that if a motor stops it must only be possible for it to be restarted manually. This is only the case where the automatic restarting of the motor may cause danger; where the motor is part of a machine that requires it to stop and start frequently, it is acceptable to allow this provided precautions are taken to prevent danger.

554.2 Heaters having immersed elements

Any element which is immersed in a liquid must have a thermal cut-out as well as a thermostat.

The thermostat is to control the temperature of the liquid and the thermal cut-out is to switch the element off and prevent danger in the event of the thermostat failing and allowing the liquid to heat manually.

There are many immersion heaters installed which do not have this safety device, and failing to look for this thermal cut-out is a common oversight during periodic inspections.

554.4 Heating conductors and cables

Where cables are used for heating, great care has to be taken to prevent the heat from the cable causing damage or fire. For this reason the regulations require us to make certain that any materials near the heating cable comply with the ignitability characteristic 'P'. For a building material to comply with this requirement tests are carried out to BS 476.

Characteristic 'P' requires that the materials are subjected to a 10-mm-long flame for a period of 10 s under specified conditions. To comply with the requirements, the flame must go out within 10 s of the removal of the flame. Plastic conduit or trunking would comply with ignitability characteristic 'P' provided it was manufactured to the required British Standard.

Where heating cables are used and are buried in soil or construction materials, they must be suitably protected to prevent any mechanical damage or corrosion. The cable must also be completely embedded in the substance which it is intended to heat as well as being installed so as not to be damaged by any movement of the material in which it is buried. As with all electrical equipment it is important to follow the manufacturer's instructions.

559.4 General requirements for outdoor lighting installations

As with all other electrical equipment which we install, luminaires must comply with British Standards and manufacturer's instructions must be followed.

Care must always be taken to ensure that where light fittings are concealed they can be accessed for maintenance such as cleaning and lamp replacement. Luminaires also produce heat and this must be taken into consideration, particularly where the luminaires are to be installed in a confined space.

Where the luminaires are connected to the fixed wiring, it is important that the connection is in compliance with all other sections of BS 7671.

Acceptable connections are:

- Plug and socket outlet
- Plug in lighting unit
- Ceiling rose (only one flexible cable per rose)
- Joint boxes
- Any other connection unit which would suitably enclose the terminals.

When connecting luminaires it is not permissible to use block connectors which are then taped up and pushed back through the ceiling. The connectors must be within a suitable box. A box could be formed of a suitable back plate and the fitting itself (Figure 5.23).

559.6.1.5 Fixing of the luminaires

The vast majority of light fittings are installed above head height, so clearly it is important that the fittings are secure. BS 7671 requires that the fixing for something as light as a pendant set has to be capable of supporting a mass of 5 kg.

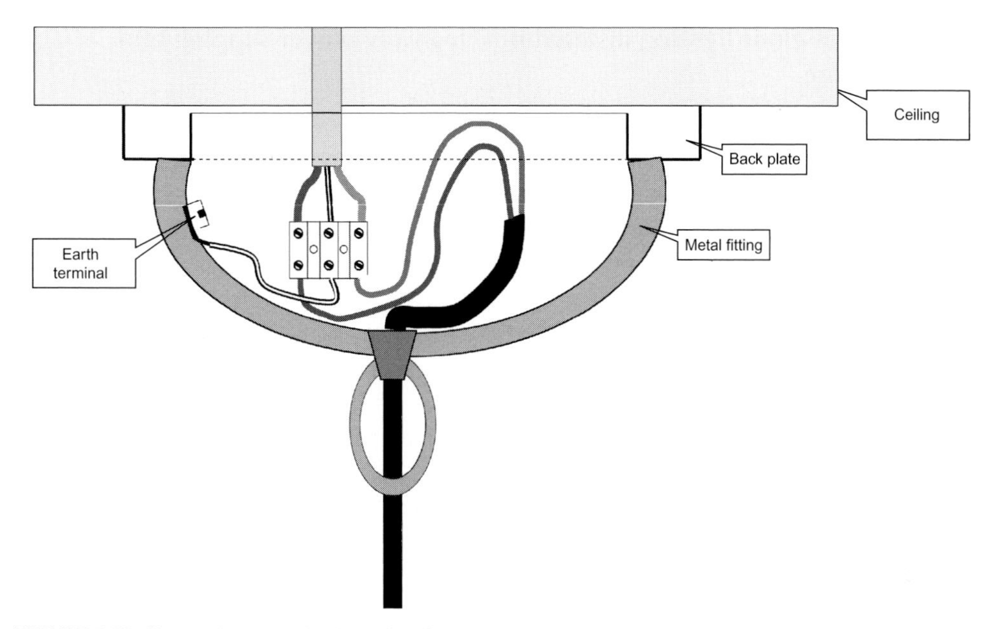

FIGURE 5.23 **Connections completely enclosed**

This requirement effectively rules out the use of plasterboard fixings unless you can be certain that the plasterboard is thick enough and very well supported (*some lampshades are very heavy*). It is always better to install the correct fixing blocks, although it is not always convenient. In situations where luminaires have a mass of greater than 5 kg, the fixing must be capable of supporting them. Where a number of luminaires is installed in suspended ceiling grids, advice should be sought as to the strength of the ceiling supports: a modular light fitting usually weighs far more than a ceiling tile which the ceiling has been designed for. It may be easier to add additional support to a ceiling than to try and support each luminaire individually.

Generally, lampholders used for luminaires will have the highest lamp wattage indicated on them: BS 7671 requires that bayonet lampholders B15 (SBC) and B22 (BC) must be able to withstand a lamp cap temperature of 210°C. This may seem complicated but in reality if we use equipment to the required British Standard we will have no problems.

All Edison screw lampholders must have the centre pin of the lampholder connected to the line and the outer terminal connected to the neutral unless the lampholders are of the all-insulated type E14 and E27.

559.6.2.2 Through wiring

Where ease of installation requires that wiring has to be run through a luminaire, care has to be taken to ensure that the heat generated by the luminaire does not cause damage to the wiring. Unless there are clear instructions available it is always wise to use heat-resistant cables.

559.10.3.1 Outdoor lighting installations, highway power supplies and street furniture

The only protective measures permitted are ADS, double or reinforced insulation and SELV.

Where ADS is used enclosures which are accessible to ordinary persons must only be accessible by the use of a tool or key.

Figure 5.24 shows the symbols used in luminaires and control gear for luminaires.

CHAPTER 56

Safety services

This section deals with the electrical supply provided to operate equipment which has been installed to enhance the safety of a building.

The supply for a safety service would normally be additional to the normal supply for the building, as in most cases the safety service would need to continue to operate after loss of the normal supply.

A safety service could include any of the following:

- Fire-alarm or fire-detection system
- Emergency lighting
- Carbon monoxide detection/alarm systems
- Smoke ventilation systems
- Safety systems.

In fact, any system which is installed and would be required to operate for reasons of safety could be classed as a safety system.

Supplies to operating theatres or areas where heat, light and sometimes sound need to be maintained to ensure the safety of persons, livestock or property would all come under this category.

These safety services need to operate unaffected by fire for as long as is possible. For this reason, the safety services need to be installed using fire-resistant cables and have suitable mechanical protection. Consideration should also be given to the siting of equipment as some buildings have water sprinkler systems which could affect the correct operation of the safety service.

Where a safety service is installed the wiring must be independent of the circuits within the building which form the normal installation. It is important that maintenance can be carried out on one system without affecting the operation of the other.

It is also a requirement that documentation, including schematic diagrams, must be provided showing:

- Full details of the electrical safety services
- All electrical control equipments
- What the final circuit does
- Any monitoring devices

Luminaire for use with a bowl

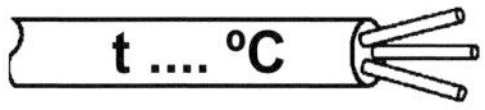

Heat resistant cable required, number of cores and temperature indicated

Luminaire may be installed on a normally flammable surface and covered with thermal insulation

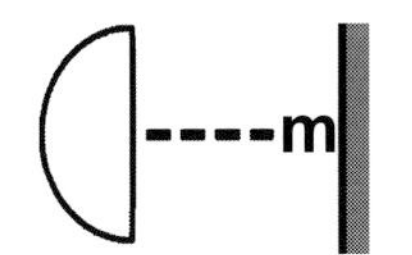

Minimum distance that the lamp may be placed from the object which it is being used to illuminate

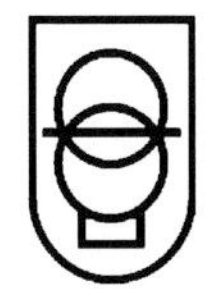

Short circuit proof transformer

Warning against the use of cool beam lamps

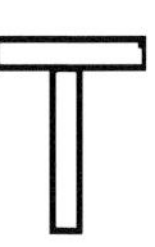

Rough service lamps

Declared temperature thermally protected ballast or transformer up to a maximum of 130°C

Class P thermally protected transformer or ballast

Luminaire intended to be used with a high pressure sodium lamp (SON) that require an external igniter

Electronic converter for an extra low voltage installation (switch mode power supply)

FIGURE 5.24 Lighting symbols

Luminaire for use with a high pressure sodium lamp (SON) having an internal starter

Luminaire suitable for mounting on a normally flammable surface

Symbol for an independent ballast

ta .... °C

Maximum ambient temperature

Suitable for mounting on non-combustible surfaces only

Luminaire with limited surface temperature

FIGURE 5.24 ***(Continued)***

- A complete list of all current-using equipment connected to the safety service; this must also show the current rating and operating time
- Operating instructions for the equipment connected to, and the safety service itself.

Safety services are often a requirement of licensing authorities, and it is usually a good idea to consult them as to their particular requirements. Often this will involve talking to the local fire officer and building control.

Chapter 35 provides information on the recognised sources for safety services. These can be:

- Cells
- Storage batteries
- Independent generators
- A separate supply provided by the supply company, possibly from a different feed so that both supplies will not fail at the same time.

In many cases, the supply to a safety service is as simple as the battery in the alarm panel or emergency light. In larger installations or where a greater load is to be connected, the safety service could be supplied by other means such as a standby generator or an uninterruptible power supply.

Part 6

INSPECTION AND TESTING

All installations have to be inspected and tested. Ideally this should be an ongoing process which starts the moment work on the installation commences, then continues for the life of the installation to ensure that it remains safe and reliable for continued service.

The EAWR 1989 states that all electrical systems should be designed, installed and maintained to be safe. Of course, the way to achieve compliance with these requirements is to carry out an initial verification on a new installation or circuit before it is put into service and then to periodically inspect the installation to ensure that it remains safe.

Building Regulations part P also requires that most electrical work carried out in the domestic sector must be certificated, and notified to the local building control.

There is often confusion over part P as many training providers advertise courses offering to make candidates part P compliant. There are no courses which will gain anybody part P compliance. There are options, however.

Reporting can be by the completion of a **building notice**, which will usually involve a fee which is set by the local authority. On completion of the building notice, work can usually commence after notifying the building control officer, who will want to ensure that on completion of the work, the person who is responsible for doing the work will be able to provide the correct type of certification. The building control officer will also, in most cases, need to look at the work at various stages as it progresses to completion.

In instances where the electrical work is part of a job which is going to be under building control, there is no requirement to complete a separate building notice. This would usually apply to new buildings, extensions or building alterations where

A Practical Guide to the 17th Edition of the Wiring Regulations. DOI: 10.1016/B978-0-08-096560-4.00006-0

planning permission and building regulation approval have been applied for and granted. In these situations, the electrician carrying out the work will need to be able to convince the building control officer that they have the required competence. There is no requirement to be qualified.

The problem with this approach is that in all cases where the electrical work is not associated with other building work the cost will be quite high. This would be uneconomic where the job is perhaps the addition of a socket outlet in a kitchen.

A far better option is to become a member of a self-registration body such as the NICEIC or NAPIT. To register with one of these organisations, you will have to pay an annual fee plus a small fee for each certificate which you complete. All that is required then is the completion of the correct certificate and the notification of the work to the registration body.

Membership is not limited to qualified electricians. The requirement is that you prove your competence to the organisation concerned. This will be achieved by a visit from an inspector who will want to look at some of your work. You will also be required to carry out a range of tests in front of the inspector and possibly answer a few questions.

There is no qualification which will gain anyone automatic membership, although in some instances a qualification may be required such as the 17th Edition Wiring Regulations (2382-10 or 20) or the City and Guilds Inspection and Testing qualification (2391-20).

Table 6.1 provides information on notifiable and non-notifiable work.

CHAPTER 61

Initial verification

On the completion of an electrical installation, it is a requirement of BS 7671 that a certificate is issued to show that the installation is compliant. The certificate that must be issued is an electrical installation certificate.

This certificate is used for a new installation, or an addition or alteration to an existing installation where the characteristics of the circuit have been altered. Certificates provided by most certification bodies include the schedule of test results and the schedule of inspections. BS 7671 shows these certificates individually (Figure 6.1).

The verification must be carried out during and on completion of the work which is being certificated and must be carried out by a competent person.

There are many types of electrical installation and the requirements for them will vary from job to job. Where relevant, the following items should be inspected to ensure that they comply with BS 7671, during erection if possible and on completion before the installation is put into service.

- Have correct erection methods been used?
- Are diagrams and instructions available where required?

TABLE 6.1 Notifiable and non-notifiable work

Work carried out	Notifiable	Non-notifiable
New installation	Yes	
Rewire	Yes	
New circuit	Yes	
Replacement circuit		Not notifiable provided the cable follows the same route as the original and the cable is the same size. It must only serve one circuit or distribution board
Replacement consumer unit	Yes	
Change of protective device type or rating	Yes	
Installation of RCD	Yes	
Work carried out in a bathroom, kitchen, swimming pool or sauna	Yes; these areas are classed as special locations	Not notifiable if only replacing an accessory or fitting
Adding an outlet to an existing circuit		No, provided they are not in a special location
Installing or upgrading protective bonding		No
Work carried out in special installations: remote buildings, gardens, solar PV, small-scale generators, extra-low-voltage lighting systems	Yes	Extra-low-voltage systems are not notifiable, provided that they are pre-assembled or individual, extra-low-voltage luminaires which are installed using individual transformers connected to existing lighting points

- Have warning and danger notices been fitted in the correct place?
- Is there suitable access to consumer units and equipment?
- Is the equipment suitable for the environment in which it has been fixed?
- Have the correct type and size of protective devices been used?
- Have 30-mA RCDs been fitted where required?
- Are socket outlets 3 m from Zone 1 when installed in a bathroom? Are they protected by a 30-mA RCD?
- Are all circuits supplying equipment in a bathroom protected by an RCD, or are the circuits SELV or PELV?
- Are the isolators and switches fitted in the correct place?
- Could the installation be damaged by work being carried out on other services or by movement due to expansion of other services?
- Are band I and II circuits separated?

ICM4

Original (To the person ordering the work)

This safety certificate is an important and valuable document which should be retained for future reference

ELECTRICAL INSTALLATION CERTIFICATE

Issued in accordance with *British Standard BS 7671- Requirements for Electrical Installations*

DETAILS OF THE CLIENT

Client / Address:

DETAILS OF THE INSTALLATION

The installation is:

Address: New

Extent of the installation covered by this certificate: An addition / An alteration

DESIGN

I/We, being the person(s) responsible for the design of the electrical installation (as indicated by my/our signature(s) below), particulars of which are described above, having exercised reasonable skill and care when carrying out the design, hereby CERTIFY that the design work for which I/we have been responsible is to the best of my/our knowledge and belief in accordance with BS 7671: amended to (date) except for the departures, if any, detailed as follows:

Details of departures from BS 7671, as amended (Regulations 120.3,120.4):

The extent of liability of the signatory/signatories is limited to the work described above as the subject of this certificate.
For the **DESIGN** of the installation: **(*Where there is divided responsibility for the design*)

Signature Date Name (CAPITALS) Designer 1

Signature Date Name (CAPITALS) ** Designer 2

CONSTRUCTION

I/We, being the person(s) responsible for the construction of the electrical installation (as indicated by my/our signature below), particulars of which are described above, having exercised reasonable skill and care when carrying out the construction, hereby CERTIFY that the construction work for which I/we have been responsible is to the best of my/our knowledge and belief in accordance with BS 7671: amended to (date) except for the the departures, if any, detailed as follows:

Details of departures from BS 7671, as amended (Regulations 120.3,120.4):

The extent of liability of the signatory is limited to the work described above as the subject of this certificate.
For the **CONSTRUCTION** of the installation:

Signature Date Name (CAPITALS) Constructor

INSPECTION AND TESTING

I/We, being the person(s) responsible for the inspection and testing of the electrical installation (as indicated by my/our signatures below), particulars of which are described above, having exercised reasonable skill and care when carrying out the inspection and testing, hereby CERTIFY that the work for which I/we have been responsible is to the best of my/our knowledge and belief in accordance with BS 7671: amended to (date) except for the departures, if any, detailed as follows:

Details of departures from BS 7671, as amended (Regulations 120.3,120.4):

The extent of liability of the signatory/signatories is limited to the work described above as the subject of this certificate.
For the **INSPECTION AND TESTING** of the installation: Reviewed by †

Signature Date Signature Date

Name (CAPITALS) Inspector Name (CAPITALS)

DESIGN, CONSTRUCTION, INSPECTION AND TESTING *

* This box to be completed only where the design, construction, inspection and testing have been the responsibility of one person.

I, being the person responsible for the design, construction, inspection and testing of the electrical installation (as indicated by my signature below), particulars of which are described above, having exercised reasonable skill and care when carrying out the design, construction, inspection and testing, hereby CERTIFY that the said work for which I have been responsible is to the best of my knowledge and belief in accordance with BS 7671, amended to (date) except for the departures, if any, detailed as follows:

Details of departures from BS 7671, as amended (Regulations 120.3, 120.4):

The extent of liability of the signatory is limited to the work described above as the subject of this certificate.
For the **DESIGN**, the **CONSTRUCTION** and the **INSPECTION AND TESTING** of the installation. Reviewed by †

Signature Date Signature Date

Name (CAPITALS) Name (CAPITALS)

† *The completed schedules of inspection and testing should preferably be reviewed by another competent person to confirm that the recorded results are consistent with electrical installation work conforming to the requirements of BS 7671*

Page 1 of

This form is based on the model shown in Appendix 6 of BS 7671: 2008.

Please see the 'Notes for Recipients' on the reverse of this page.

ICM4/1

FIGURE 6.1 ICM4 Electrical Installation Certificate

NOTES FOR RECIPIENT

THIS SAFETY CERTIFICATE IS AN IMPORTANT AND VALUABLE DOCUMENT WHICH SHOULD BE RETAINED FOR FUTURE REFERENCE

This safety certificate has been issued to confirm that the electrical installation work to which it relates has been designed, constructed, inspected, tested and verified in accordance with the national standard for the safety of electrical installations, British Standard 7671 (as amended) - *Requirements for Electrical Installations.*

Where, as will often be the case, the installation incorporates a residual current device (RCD), there should be a notice at or near the main switchboard or consumer unit stating that the device should be tested at quarterly intervals. For safety reasons, it is important that you carry out the test regularly.

Also for safety reasons, the complete electrical installation will need to be inspected and tested at appropriate intervals by a competent person. The maximum interval recommended before the next inspection is stated on Page 2 under *Next Inspection*. There should be a notice at or near the main switchboard or consumer unit indicating when the inspection of the installation is next due.

This report is intended for use by electrical contractors not enrolled with

NICEIC or by NICEIC Approved Contractors working outside the scope of their enrolment. The certificate consists of at least five numbered pages.

For installations having more than one distribution board or more circuits than can be recorded on pages 4 and 5, one or more additional *Schedules of Circuit Details for the Installation*, and *Schedules of Test Results for the Installation* (pages 6 and 7 onwards) should form part of the certificate.

This certificate is intended to be issued only for a new electrical installation or for new work associated with an alteration or addition to an existing installation. It should not have been issued for the inspection of an existing electrical installation. A 'Periodic Inspection Report' should be issued for such a periodic inspection.

You should have received the certificate marked 'Original' and the electrical contractor should have retained the certificate marked 'Duplicate'.

If you were the person ordering the work, but not the user of the installation, you should pass this certificate, or a full copy of it including these notes, the schedules and additional pages (if any), immediately to the user.

The 'Original' certificate should be retained in a safe place and shown to any person inspecting or undertaking further work on the electrical installation in the future. If you later vacate the property, this certificate will demonstrate to the new user that the electrical installation complied with the requirements of the national electrical safety standard at the time the certificate was issued.

Page 1 of this certificate provides details of the electrical installation, together with the name(s) and signature(s) of the person(s) certifying the three elements of installation work: design, construction and inspection and testing. Page 2 identifies the organisation(s) responsible for the work certified by their representative(s).

Certification for inspection and testing provides an assurance that the electrical installation work has been fully inspected and tested, and that the electrical work has been carried out in accordance with the requirements of BS 7671 (except for any departures sanctioned by the designer and recorded in the appropriate box(es) of the certificate).

If wiring alterations or additions are made to an installation such that wiring colours to two versions of BS 7671 exist, a warning notice should have been affixed at or near the appropriate consumer unit.

continued on the reverse of page 2

ICM4/1&2B

FIGURE 6.1 ***(Continued)***

ICM4

Original (To the person ordering the work)

PARTICULARS OF THE ORGANISATION(S) RESPONSIBLE FOR THE ELECTRICAL INSTALLATION

DESIGN (1) Organisation †

Address:

Postcode

DESIGN (2) Organisation †

Address:

Postcode

† CONSTRUCTION Organisation

Address:

Postcode

INSPECTION AND TESTING Organisation †

Address:

Postcode

SUPPLY CHARACTERISTICS AND EARTHING ARRANGEMENTS

Tick boxes and enter details, as appropriate

✣ System Type(s)	✣ Number and Type of Live Conductors			Nature of Supply Parameters		✣ Characteristics of Primary Supply Overcurrent Protective Device(s)
TN-S	a.c.		d.c.	Nominal voltage(s): $U^{(1)}$ V $U_0^{(1)}$ V		
TN-C-S	1-phase (2 wire)	1-phase (3 wire)	2 pole	Nominal frequency, $f^{(1)}$ Hz	*Notes:* *(1) by enquiry*	BS(EN)
TN-C	2-phase (3 wire)		3-pole	Prospective fault current, $I_{pf}^{(2)(3)}$ kA	*(2) by enquiry or by measurement*	Type
TT	3-phase (3 wire)	3-phase (4 wire)	other	External earth fault loop impedance, $Z_e^{(2)(3)}$ Ω	*(3) where more than one supply, record the higher or highest values*	Rated current A
IT	Other	Please state		Number of supplies		Short-circuit capacity kA

PARTICULARS OF INSTALLATION AT THE ORIGIN

Tick boxes and enter details, as appropriate

✣ Means of Earthing	Details of Installation Earth Electrode (where applicable)	
Distributor's facility:	Type: (eg rod(s), tape etc)	Location:
Installation earth electrode:	Electrode resistance, R_A: (Ω)	Method of measurement:

✣ Main Switch or Circuit-Breaker

* (applicable only where an RCD is suitable and is used as a main circuit-breaker)

Maximum Demand (Load): kVA / Amps *Delete as appropriate

Protective measure(s) against electric shock:

Main Switch or Circuit-Breaker		Earthing conductor	Main protective bonding conductors	Bonding of extraneous-conductive-parts (✓)	
Type: BS(EN)	Voltage rating V				
No of Poles	Rated current, I_n A	Conductor material	Conductor material	Water service	Gas service
Supply conductors: material	RCD operating current, $I_{\Delta n}$* mA	Conductor csa mm²	Conductor csa mm²	Oil service	Structural steel
Supply conductors: csa mm²	RCD operating time (at $I_{\Delta n}$)* ms	Continuity check (✓)	Continuity check (✓)	Lightning protection	Other incoming service(s)

Earthing and Protective Bonding Conductors

COMMENTS ON EXISTING INSTALLATION

In the case of an alteration or additions see Section 633

Note: Enter 'NONE' or, where appropriate, the page number(s) of additional page(s) of comments on the existing installation.

NEXT INSPECTION

§ *Enter interval in terms of years, months or weeks, as appropriate*

I/We, the designer(s), RECOMMEND that this installation is further inspected and tested after an interval of not more than §

† *Where the electrical contractor responsible for the construction of the electrical installation has also been responsible for the design **and** the inspection and testing of that installation, the 'Particulars of the Organisation Responsible for the Electrical Installation' may be recorded only in the section entitled 'CONSTRUCTION'.*

✣ *Where a number of sources are available to supply the installation, and where the data given for the primary source may differ from other sources, a separate sheet must be provided which identifies the relevant information relating to each additional source.*

Page 2 of

Please see the 'Notes for Recipients' on the reverse of this page.

ICM4/3

FIGURE 6.1 ***(Continued)***

NOTES FOR RECIPIENT
(continued from the reverse of page 1)

Where responsibility for the *design*, the *construction* and the *inspection and testing* of the electrical work is divided between the electrical contractor and one or more other bodies, the division of responsibility should have been established and agreed before commencement of the work. In such a case, the absence of certification for the *construction*, or the *inspection and testing* elements of the work would render the certificate invalid. If the *design* section of the certificate has not been completed, you should question why those responsible for the design have not certified that this important element of the work is in accordance with the national electrical safety standard.

All unshaded boxes should have been completed either by insertion of the relevant details or by entering 'N/A', meaning 'Not Applicable', where appropriate.

Where the electrical work to which this certificate relates includes the installation of a fire alarm system and/or an emergency lighting system (or a part of such systems) in accordance with British Standards BS 5839 and BS 5266 respectively, this electrical safety certificate should be accompanied by a separate certificate or certificates as prescribed by those standards.

Where the installation can be supplied by more than one source, such as the public supply and a standby generator, the number of sources should have been recorded in the box entitled Number of Supplies, under the general heading *Supply Characteristics and Earthing Arrangements* on page 2 of the certificate, and the *Schedule of Test Results* compiled accordingly. Where a number of sources are available to supply the installation, and where the data given for the primary source may differ from other sources, an additional page should have been provided which gives the relevant information relating to each additional source, and to the associated earthing arrangements and main switchgear.

ICM4/3&4B

FIGURE 6.1 ***(Continued)***

ICM4

Original (To the person ordering the work)

SCHEDULE OF ITEMS INSPECTED

† See note below

PROTECTIVE MEASURES AGAINST ELECTRIC SHOCK

Basic and fault protection

Extra low voltage

- [ ] SELV
- [ ] PELV

Double or reinforced insulation

- [ ] Double or Reinforced Insulation

Basic protection

- [ ] Insulation of live parts
- [ ] Barriers or enclosures
- [ ] Obstacles **
- [ ] Placing out of reach **

Fault protection

Automatic disconnection of supply

- [ ] Presence of earthing conductor
- [ ] Presence of circuit protective conductors
- [ ] Presence of main protective bonding conductors
- [ ] Presence of earthing arrangements for combined protective and functional purposes
- [ ] Presence of adequate arrangements for alternative source(s), where applicable
- [ ] FELV
- [ ] Choice and setting of protective and monitoring devices (for fault protection and/or overcurrent protection)

Non-conducting location **

- [ ] Absence of protective conductors

Earth-free equipotential bonding **

- [ ] Presence of earth-free equipotential bonding

Electrical separation

- [ ] For **one** item of current-using equipment
- [ ] For **more** than one item of current-using equipment **

Additional protection

- [ ] Presence of residual current device(s)
- [ ] Presence of supplementary bonding conductors

***** For use in controlled supervised/conditions only***

Prevention of mutual detrimental influence

- [ ] Proximity of non-electrical services and other influences
- [ ] Segregation of Band I and Band II circuits or Band II insulation used
- [ ] Segregation of Safety Circuits

Identification

- [ ] Presence of diagrams, instructions, circuit charts and similar information
- [ ] Presence of danger notices and other warning notices
- [ ] Labelling of protective devices, switches and terminals
- [ ] Identification of conductors

Cables and Conductors

- [ ] Selection of conductors for current carrying capacity and voltage drop
- [ ] Erection methods
- [ ] Routing of cables in prescribed zones
- [ ] Cables incorporating earthed armour or sheath or run in an earthed wiring system, or otherwise protected against nails, screws and the like
- [ ] Additional protection by 30mA RCD for cables concealed in walls (where required, in premises not under the supervision of skilled or instructed persons)
- [ ] Connection of conductors
- [ ] Presence of fire barriers, suitable seals and protection against thermal effects

General

- [ ] Presence and correct location of appropriate devices for isolation and switching
- [ ] Adequacy of access to switchgear and other equipment
- [ ] Particular protective measures for special installations and locations
- [ ] Connection of single-pole devices for protection or switching in line conductors only
- [ ] Correct connection of accessories and equipment
- [ ] Presence of undervoltage protective devices
- [ ] Selection of equipment and protective measures appropriate to external influences
- [ ] Selection of appropriate functional switching devices

SCHEDULE OF ITEMS TESTED

† See note below

- [ ] External earth fault loop impedance, Z_e
- [ ] Installation earth electrode resistance, R_A
- [ ] Continuity of protective conductors
- [ ] Continuity of ring final circuit conductors
- [ ] Insulation resistance between live conductors
- [ ] Insulation resistance between live conductors and Earth
- [ ] Protection by SELV, PELV or by electrical separation
- [ ] Basic protection by barrier or enclosure provided during erection
- [ ] Insulation of non-conducting floors or walls
- [ ] Polarity
- [ ] Earth fault loop impedance, Z_s
- [ ] Verification of phase sequence
- [ ] Operation of residual current devices
- [ ] Functional testing of assemblies
- [ ] Verification of voltage drop

SCHEDULE OF ADDITIONAL RECORDS* (See attached schedule)

Page No(s)

Note: Additional page(s) must be identified by the Electrical Installation Certificate serial number and page number(s).

† ***All boxes must be completed.*** *'✓' indicates that an inspection or a test was carried out and that the result was **satisfactory**. 'N/A' indicates that an inspection or test was **not applicable** to the particular installation.*

Page 3 of

* *Where the electrical work to which this certificate relates includes the installation of a fire alarm system and/or an emergency lighting system (or a part of such systems), this electrical safety certificate should be accompanied by the particular certificate(s) for the system(s).*

This form is based on the model shown in Appendix 6 of BS 7671: 2008.

ICM4/5

FIGURE 6.1 ***(Continued)***

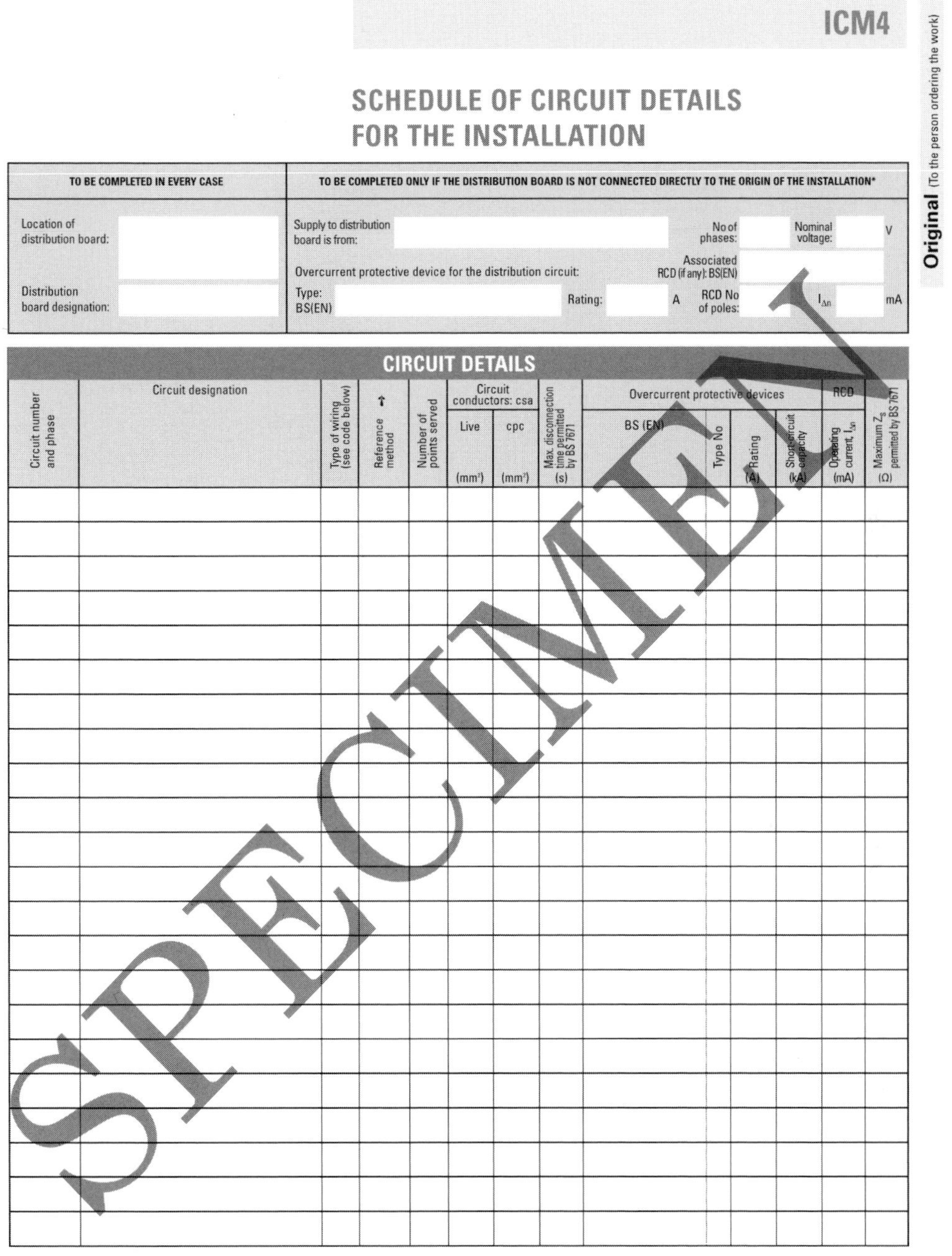

ICM4

Original (To the person ordering the work)

SCHEDULE OF CIRCUIT DETAILS FOR THE INSTALLATION

TO BE COMPLETED IN EVERY CASE	TO BE COMPLETED ONLY IF THE DISTRIBUTION BOARD IS NOT CONNECTED DIRECTLY TO THE ORIGIN OF THE INSTALLATION*
Location of distribution board:	Supply to distribution board is from: No of phases: Nominal voltage: V
	Overcurrent protective device for the distribution circuit: Associated RCD (if any): BS(EN)
Distribution board designation:	Type: BS(EN) Rating: A RCD No of poles: $I_{\Delta n}$ mA

CIRCUIT DETAILS

Circuit number and phase	Circuit designation	Type of wiring (see code below)	Reference method †	Number of points served	Circuit conductors: csa Live (mm²)	Circuit conductors: csa cpc (mm²)	Max. disconnection time permitted by BS 7671 (s)	Overcurrent protective devices BS (EN)	Overcurrent protective devices Type No	Overcurrent protective devices Rating (A)	Overcurrent protective devices Short-circuit capacity (kA)	RCD Operating current, $I_{\Delta n}$ (mA)	Maximum Z_s permitted by BS 7671 (Ω)

† *See Table 4A2 of Appendix 4 of BS 7671: 2008*

CODES FOR TYPE OF WIRING								
A	B	C	D	E	F	G	H	O (Other - please state)
PVC/PVC cables	PVC cables in metallic conduit	PVC cables in non-metallic conduit	PVC cables in metallic trunking	PVC cables in non-metallic trunking	PVC/SWA cables	XLPE/SWA cables	Mineral-insulated cables	

Page 4 of

** In such cases, details of the distribution (sub-main) circuit(s), together with the test results for the circuit(s), must also be provided on continuation schedules.*

This form is based on the model shown in Appendix 6 of BS 7671: 2008.

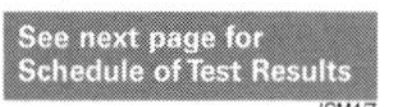

See next page for Schedule of Test Results

ICM4/7

FIGURE 6.1 (*Continued*)

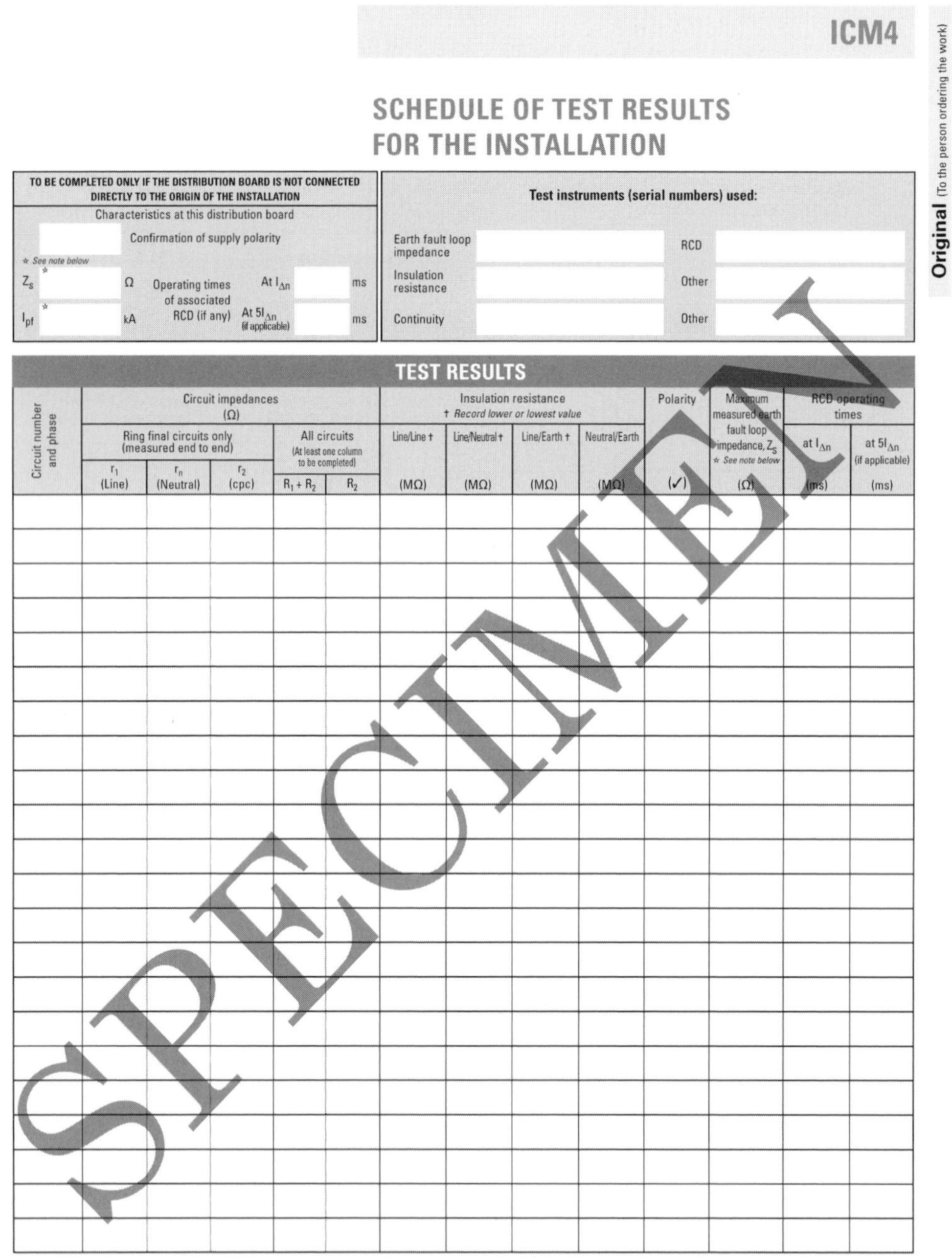

ICM4

Original (To the person ordering the work)

SCHEDULE OF TEST RESULTS FOR THE INSTALLATION

TO BE COMPLETED ONLY IF THE DISTRIBUTION BOARD IS NOT CONNECTED DIRECTLY TO THE ORIGIN OF THE INSTALLATION

Characteristics at this distribution board

Confirmation of supply polarity

* See note below

Z_s * Ω Operating times of associated RCD (if any) At $I_{\Delta n}$ ms

I_{pf} * kA At $5I_{\Delta n}$ (if applicable) ms

Test instruments (serial numbers) used:

Earth fault loop impedance		RCD	
Insulation resistance		Other	
Continuity		Other	

TEST RESULTS

Circuit number and phase	Circuit impedances (Ω)					Insulation resistance † Record lower or lowest value				Polarity	Maximum measured earth fault loop impedance, Z_s * See note below	RCD operating times	
	Ring final circuits only (measured end to end)			All circuits (At least one column to be completed)		Line/Line †	Line/Neutral †	Line/Earth †	Neutral/Earth			at $I_{\Delta n}$	at $5I_{\Delta n}$ (if applicable)
	r_1 (Line)	r_n (Neutral)	r_2 (cpc)	$R_1 + R_2$	R_2	(MΩ)	(MΩ)	(MΩ)	(MΩ)	(✓)	(Ω)	(ms)	(ms)

* *Note: Where the installation can be supplied by more than one source, such as a primary source (eg public supply) and a secondary source (eg standby generator), the higher or highest values must be recorded.*

TESTED BY

Signature:		Position:	
Name: (CAPITALS)		Date of testing:	

Page 5 of

This form is based on the model shown in Appendix 6 of BS 7671: 2008.

See previous page for Circuit Details

ICM4/9

FIGURE 6.1 *(Continued)*

- Is basic protection in place?
- Are the requirements for ADS in place, where required?
- Have the test results been compared to the relevant criteria to ensure the correct disconnection in the event of a fault?
- Are fire barriers in place where required?
- Are the cables routed in safe zones? If not, are they protected against mechanical damage?
- Are the correct size cables being used taking into account voltage drop and current carrying requirements?
- Are protective devices and single pole switches connected in the phase conductor?
- Are the circuits identified?
- Have the conductors been connected correctly?

This list is not exhaustive and depending on the type of installation other items may need to be inspected.

A schedule of inspections must be completed for the installation as this will form part of the documentation which will be handed to the client on the completion of the work.

During the initial verification, each circuit must be tested. This will require the use of the correct type of testing equipment which is detailed later in this book.

It is important for safety reasons that during the initial verification, the testing procedure is carried out in the correct sequence as stated in guidance note 3 of BS 7671.

Sequence of tests

- Continuity of protective conductors, including main and supplementary bonding.
- Continuity of ring final circuit conductors.
- Insulation resistance.
- Protection of SELV and PELV.
- Protection by electrical separation.
- Protection by barriers and enclosures.
- Insulation of non-conducting floors.
- Dead polarity of each circuit.
- Live polarity of supply.
- Earth electrode resistance (Z_e).
- Earth fault loop impedance (Z_e) (Z_s).
- Additional protection.
- PFC.
- Check of phase sequence.
- Functional testing.
- Verification of voltage drop.

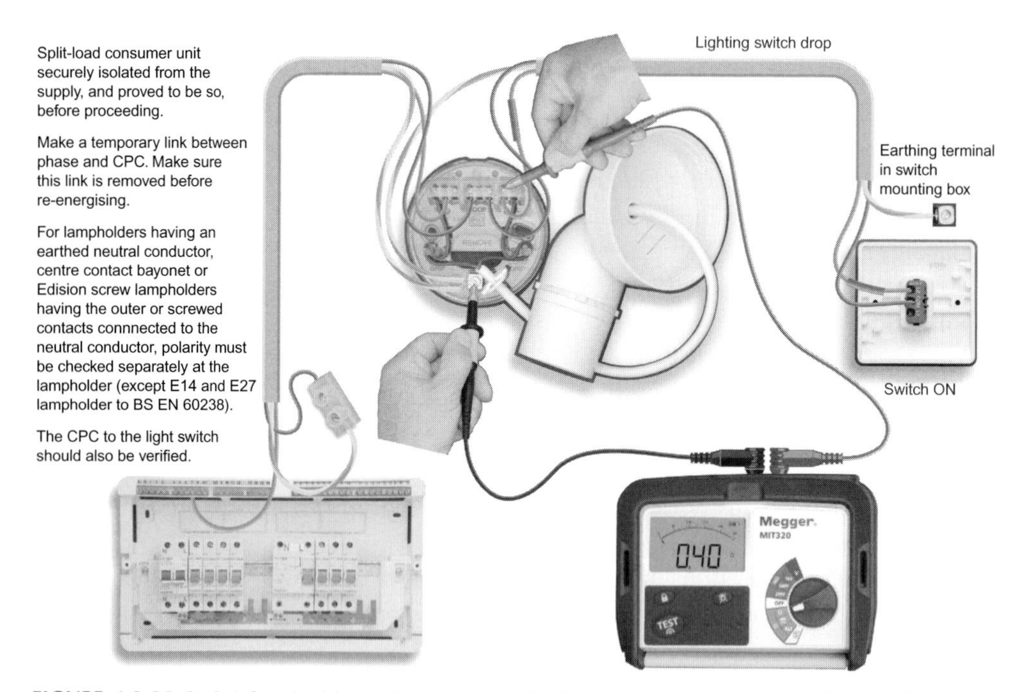

FIGURE 6.2 **Method 1 for checking polarity and continuity, and measuring $R_1 + R_2$ for compliance with regulations 612.2 and 612.6**

The results of these tests must be recorded onto a schedule of test results and compared with the relevant criteria as set out in BS 7671. If any of the results are unsatisfactory an electrical installation certificate must not be issued until the non-compliance is rectified.

Where tests are to be carried out, it is a requirement that the instruments used are compliant with BS 7671.

Continuity of protective conductors and the ring circuit test must be carried out using a low-resistance ohm-meter. For the continuity of a CPC, in most instances method 1 would be the most practical as it would give you the values of $R_1 + R_2$. This method is often called the $R_1 + R_2$ method (Figure 6.2).

The instrument must have a no-load voltage of between 4 and 24 V d.c. and deliver a minimum short circuit current of 200 mA.

For confirmation of bonding continuity method 2 should be used (Figure 6.3).

The same type of instrument should be used for a ring final test.

Insulation resistance test can be carried out on a complete installation, a section of the installation or a single circuit, whichever is the most suitable. Figure 6.3 shows the test being carried out between live conductors and earth on a complete installation (Figure 6.4).

The instrument must be capable of delivering a current of 1 mA at 250 or 500 V and where the installation or circuit voltage is above 500 V the same current must be provided at 1000 V.

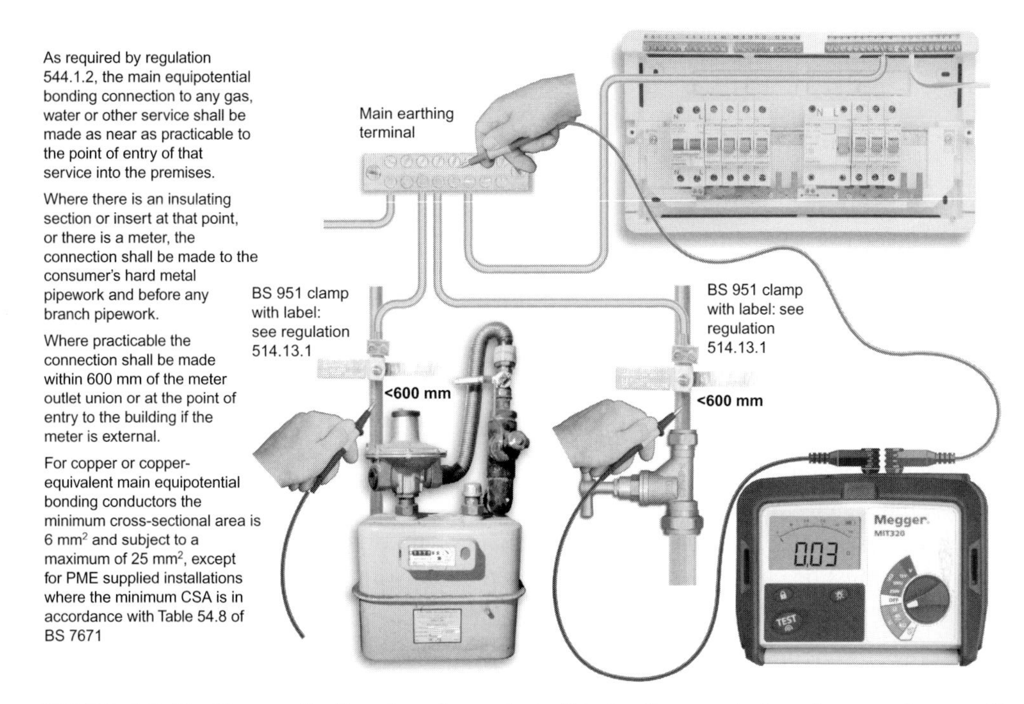

FIGURE 6.3 Checking continuity of main equipotential bonding conductors for compliance with regulation 612.2

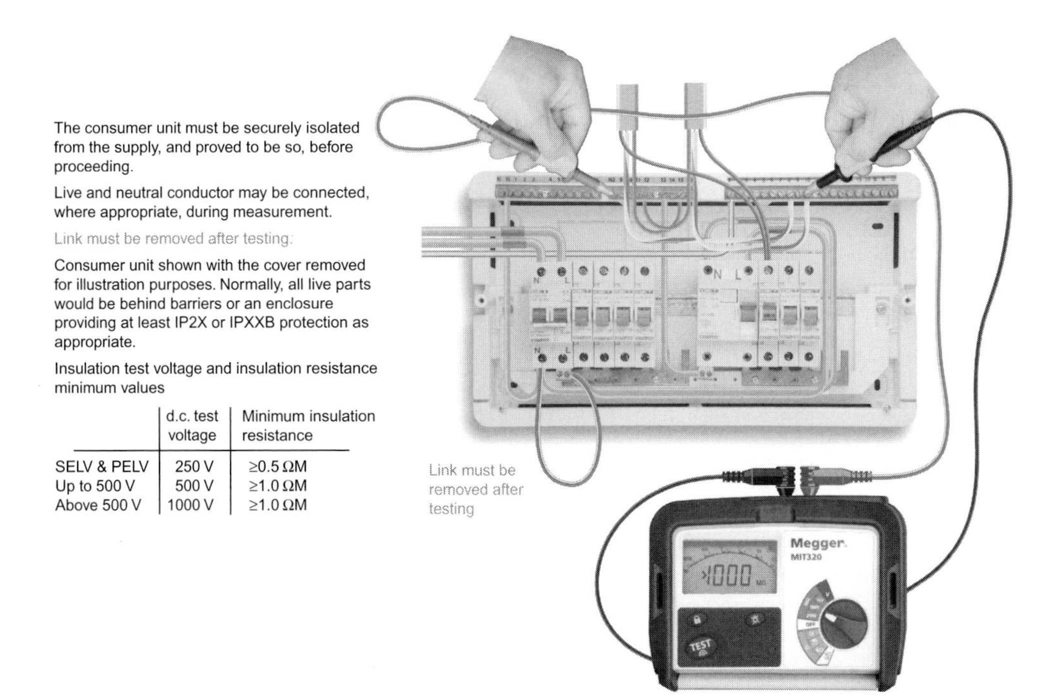

	d.c. test voltage	Minimum insulation resistance
SELV & PELV	250 V	≥0.5 ΩM
Up to 500 V	500 V	≥1.0 ΩM
Above 500 V	1000 V	≥1.0 ΩM

FIGURE 6.4 Checking insulation resistance between live conductors and protective conductor connected to the earthing arrangement for compliance with regulation 612.3.1

TABLE 6.2 Insulation resistance

Circuit nominal voltage	Test voltage	Minimum insulation resistance
SELV and PELV	250 V d.c.	≥0.5 MΩ
Low voltage between 50 and 500V	500 V d.c.	≥1 MΩ
Low voltage above 500V	1000 V d.c.	≥1 MΩ

The whole installation with all circuits combined must comply with the values given in table 5.4 (Table 6.2).

Of course, when carrying out these tests all equipment must be disconnected to avoid damage and false readings.

It is quite acceptable to test and record each circuit individually; however, a calculation must be carried out to ensure that the total value of insulation resistance will not be less than the requirement of table 5.4.

Example

An installation has six circuits with the following IR values recorded:

Circuit 1 ≥ 200 MΩ
Circuit 2 120 MΩ
Circuit 3 85 MΩ
Circuit 4 70 MΩ
Circuit 5 ≥ 200 MΩ
Circuit 6 10 MΩ

The calculation for the total is:

$$\frac{1}{R_1}+\frac{1}{R_2}+\frac{1}{R_3}+\frac{1}{R_4}+\frac{1}{R_5}+\frac{1}{R_6}=\frac{1}{R_t}=R$$

$$\frac{1}{200}+\frac{1}{120}+\frac{1}{85}+\frac{1}{70}+\frac{1}{200}+\frac{1}{10}=\frac{1}{0.144}=6.92\text{M}\Omega$$

The simplest way to carry out this calculation is to use a calculator as follows:

$$200x^{-1}+120x^{-1}+85x^{-1}+70x^{-1}+200x^{-1}+10x^{-1}=x^{-1}=6.92\text{M}\Omega$$

This installation would satisfy the requirements of BS 7671 as the insulation resistance of the complete installation measured as a whole is greater than 1 MΩ.

Where an installation has items of equipment such as surge protection, or electronic equipment which cannot be disconnected but may be damaged by a 500-V test, it is acceptable to use a test voltage of 250 V d.c. The measured value must be 1 MΩ or greater.

Protection by SELV, PELV or electrical separation

Before carrying out this test a visual check is required to ensure that the transformer for the circuit is compliant with the required British Standard (BS EN 61558-2-6).

The test will require the use of an insulation resistance test instrument.

SELV

A test must be carried out at 500 V d.c. between the live conductors on the primary side of the transformer and the live conductors on the secondary side of the transformer. The measured value must be greater than 1 MΩ.

A second test is then carried out at 500 V d.c. between all live conductors and the earth of the primary circuit (there must be no earth in the secondary circuit). The measured values must be greater than 1 MΩ.

PELV

A test must be carried out at 500 V d.c. between the live conductors on the primary side of the transformer and the live conductors on the secondary side of the transformer. The measured value must be greater than 1 MΩ.

A second test must be carried out at 500 V d.c. between all of the live conductors on the primary and secondary sides of the transformer and earth (PELV has an earthed secondary side). The measured value must be greater than 1 MΩ.

Electrical separation

The transformer used for this type of protection will have the same voltage on the primary and secondary sides of the transformer. The primary side will be earthed and the secondary will not. A test must be carried out at 500 V d.c. between the live conductors on the primary side and the live conductors on the secondary side. The measured value must be greater than 1 MΩ.

A second test is then carried out at 500 V d.c. between all live conductors and earth. The measured value must be greater than 1 MΩ.

Polarity

These tests are carried out using a low-resistance ohm-meter, or often a visual inspection will confirm polarity (at a socket outlet for example).

This test is carried out to check that all single pole switches and protective devices are connected into the line conductors of the circuit, all of the accessories are correctly connected, and ES lampholders have the centre pin connected to the line conductor. An exception is made for E14 and E27 as this type of lampholder is all insulated.

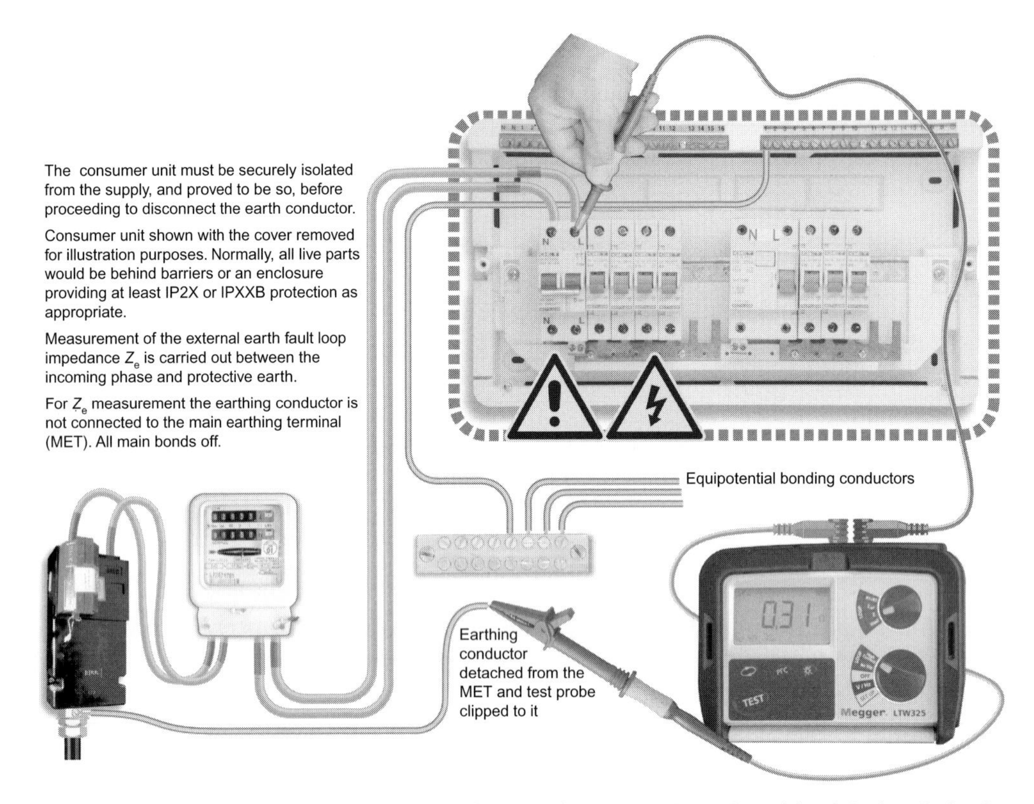

FIGURE 6.5 Measurement of the external earth fault loop impedance, Z_e, at the origin of the installation for compliance with regulation 612.9

Earth fault loop impedance

These are live tests and great care should be taken when carrying them out.

There are two measurements for earth loop impedance; one is for external loop impedance (Z_e) and the other is for the circuit loop impedance (Z_S).

Both of these tests are carried out using an earth fault loop impedance test instrument with leads and probes compliant to GS 38. The instrument uses a test current of up to 25 A for a maximum duration of 40 ms.

Z_e, external earth loop impedance

The installation must be isolated and the earthing conductor disconnected. Connect one lead to the disconnected earthing conductor and then insert a probe into the terminal of the incoming line. The measured value will be Z_e (Figure 6.5).

If the instrument has three leads then the third lead must be connected to the incoming neutral of the supply.

For obvious reasons, the earthing conductor must be reconnected before re-energising the supply to the installation.

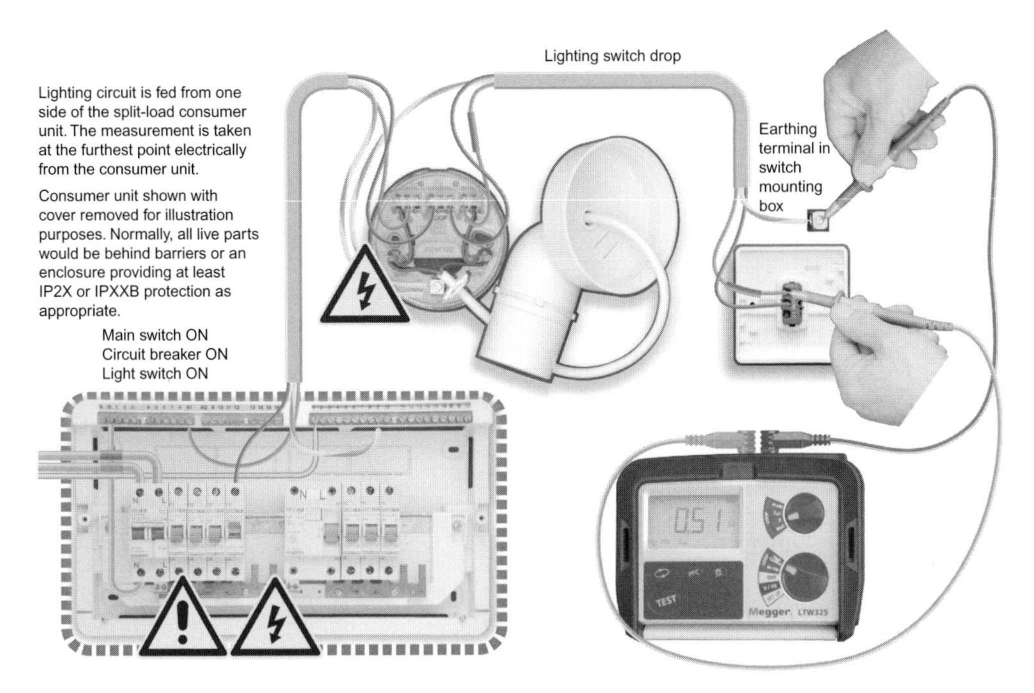

FIGURE 6.6 Measurement of earth fault loop impedance, Z_s, at lighting points for compliance with regulation 612.9

This test can also be used to measure the resistance of an earth electrode on most installations using a TT supply system.

Z_s, circuit earth fault loop impedance

This is a reasonably simple procedure but care must be taken as it is a live test (Figure 6.6).

For socket outlet circuits, all that is required is that the instrument is plugged into the socked using the lead supplied and the result recorded.

Where the circuit has no socket outlets the instrument has to be connected to the exposed terminals of the accessories on the circuit being tested, all points must be tested and the highest test result recorded as Z_s for the circuit.

The first check to be made is to look at the test result sheet and add the measured Z_e to the recorded $R_1 + R_2$ value, then compare the total value with the measured Z_s. If it is equal to or lower then all is fine; if it is higher then it may be that there is a loose connection and further investigation is required. The measured value will often be less than the calculated value due to the presence of parallel paths.

The measured Z_s must be compared to the maximum value of Z_s. This is to ensure that it complies with the requirements for the circuit disconnection time. The easiest method is to use the values given in the on-site guide or guidance note 3 as these values have been corrected to allow for conductor temperature and conductor

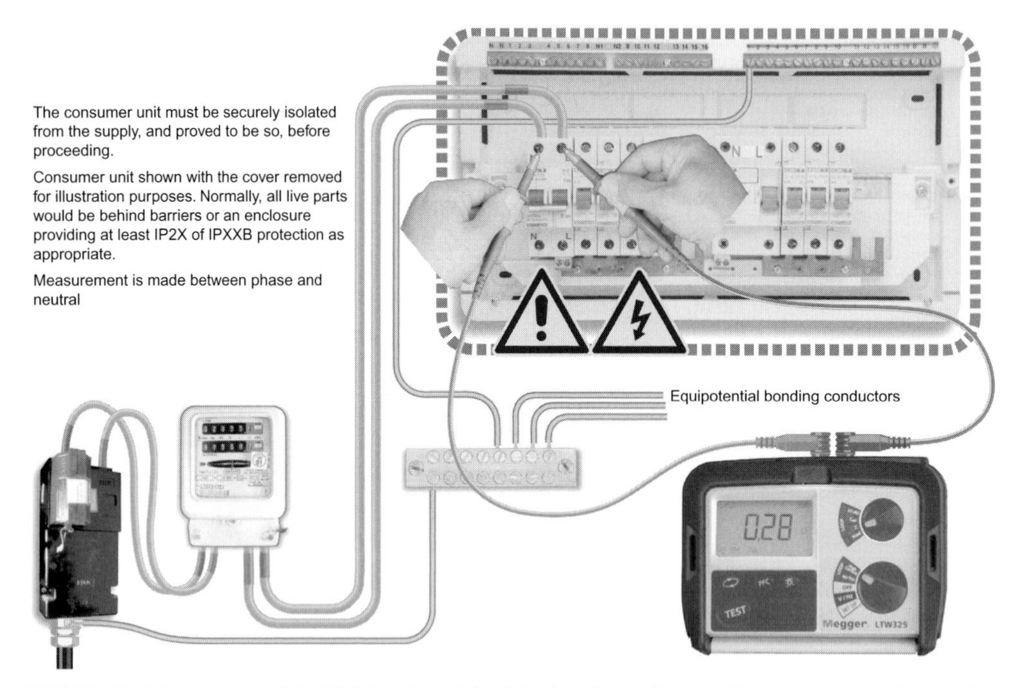

FIGURE 6.7 **Measurement of the PSCC at the origin of the installation for compliance with regulation 612.12**

operating temperature. Another method is to use the rule of thumb as described in appendix 4 of BS 7671.

As an example, let us say the circuit has a measured Z_s of 0.51 Ω and that it is protected by a type C 32 A BS EN 60898 circuit breaker. If we look in table 41.3, we can see that the maximum permissible Z_s for this circuit is 0.72 Ω.

We need to multiply the maximum permissible value by 0.8: $0.72 \times 0.8 = 0.57\,\Omega$. This now becomes the maximum permissible value. As the measured value is less than this, the circuit is acceptable.

Prospective fault current

This test uses a PFC test instrument, which is usually the same instrument used for an earth fault loop impedance but on a different setting. The value will be given in kA instead of Ω (Figure 6.7).

This test is carried out at the origin of the supply to measure the highest current that could flow in the event of a fault between live conductors.

Where possible the installation should be isolated for safety reasons. The leads of the instrument must be connected to the line and neutral of the incoming supply and the value recorded.

Where the supply is three phase, the value of current that could flow between line conductors must be measured. Some instruments will be capable of measuring this

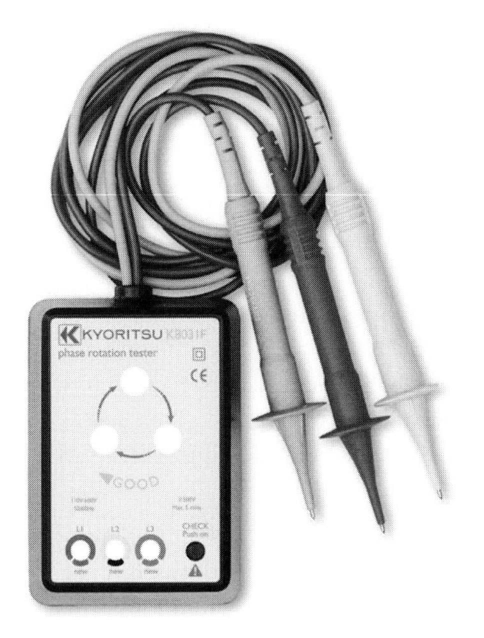

FIGURE 6.8 **Phase rotation test instrument**

value, but in instances where the instrument cannot measure between line conductors, or even if you are unsure, it is perfectly acceptable to double the value measured between line and neutral.

The object of this test is to gain a value which we can compare with installed equipment to ensure that it will not cause a danger in the event of a fault. This is particularly important for the selection of protective devices. Usually, it is sufficient to carry out this test at the origin only; however, where different types of circuit breakers are to be used within an installation the measurement can be carried out at each circuit board.

Check of phase rotation

This test is carried out to ensure that the phase sequence at each distribution board is the same as it is at the origin of the supply.

The instrument used for this test is a phase rotation test instrument; this could be an indicator lamp or of rotating disc type, and many multifunctional instruments are also capable of this test (Figure 6.8).

A lead must be connected to each incoming line at the incoming supply and then the test repeated at each section of the installation where the sequence may have been altered.

Even where this test proves that the sequence is correct it is a good idea, where possible, to run any motors before they are connected to the machinery that they are driving. Just ensure that they have the correct required rotation.

Functional testing

This involves operating any switch or circuit breaker along with any control equipment to ensure that they are all working correctly and that they are all correctly fitted and adjusted.

Functional testing will also involve the testing of any RCD which has been installed.

It is important to ensure that for safety reasons the earth fault loop impedance test is carried out before this test, as an earth must be present.

For all RCDs and RCBOs up to and including 30 mA, the test must be carried out at 0° and 180°.

$(1/2) \times I_{\Delta n}$: the device must not operate.
$1 \times I_{\Delta n}$: the device must operate within 200 ms if it is a BS type or 300 ms if it is a BS EN type.
$5 \times I_{\Delta n}$: the device must operate within 40 ms.

All of these tests must be carried out on the positive and the negative side of the half cycle and the highest value recorded.

On completion of the test sequence, the test button of the RCD should be operated to prove the operation of the test button. A check should also be made to ensure that there is a test notice sited near the RCD.

Note: When periodically testing an RCD with a meter, it is often advisable to undertake the $5 \times I_{\Delta n}$ *test first. If the device fails this test, it is a good indication that the quarterly test button check has not been carried out and that dirt or dust on the contacts may be causing a slower operation than normal. Before rejecting the device completely, the test should be carried out twice more, as it is quite usual for the device to operate properly after these two subsequent operations. In the same way as when testing the device initially, the outgoing or load cables should be removed before the test is carried out.*

To ensure that the device continues to operate within the defined parameters, it is important that the device is tripped periodically by operating the test button. This should be done at regular intervals not exceeding 3 months.

Verification of voltage drop will not normally be required on an initial verification as the voltage drop should have been calculated at the design stage of the installation.

The results of all of the tests carried out must be recorded on a schedule of test results (Figure 6.1, page 122).

A schedule of inspection must also be carried out (Figure 6.9). Each box must be filled marked with either a ✓ or N/A. Where this certificate is used for initial verification it must have no ✗ in any box as the certificates for an initial verification can only be issued on a satisfactory installation.

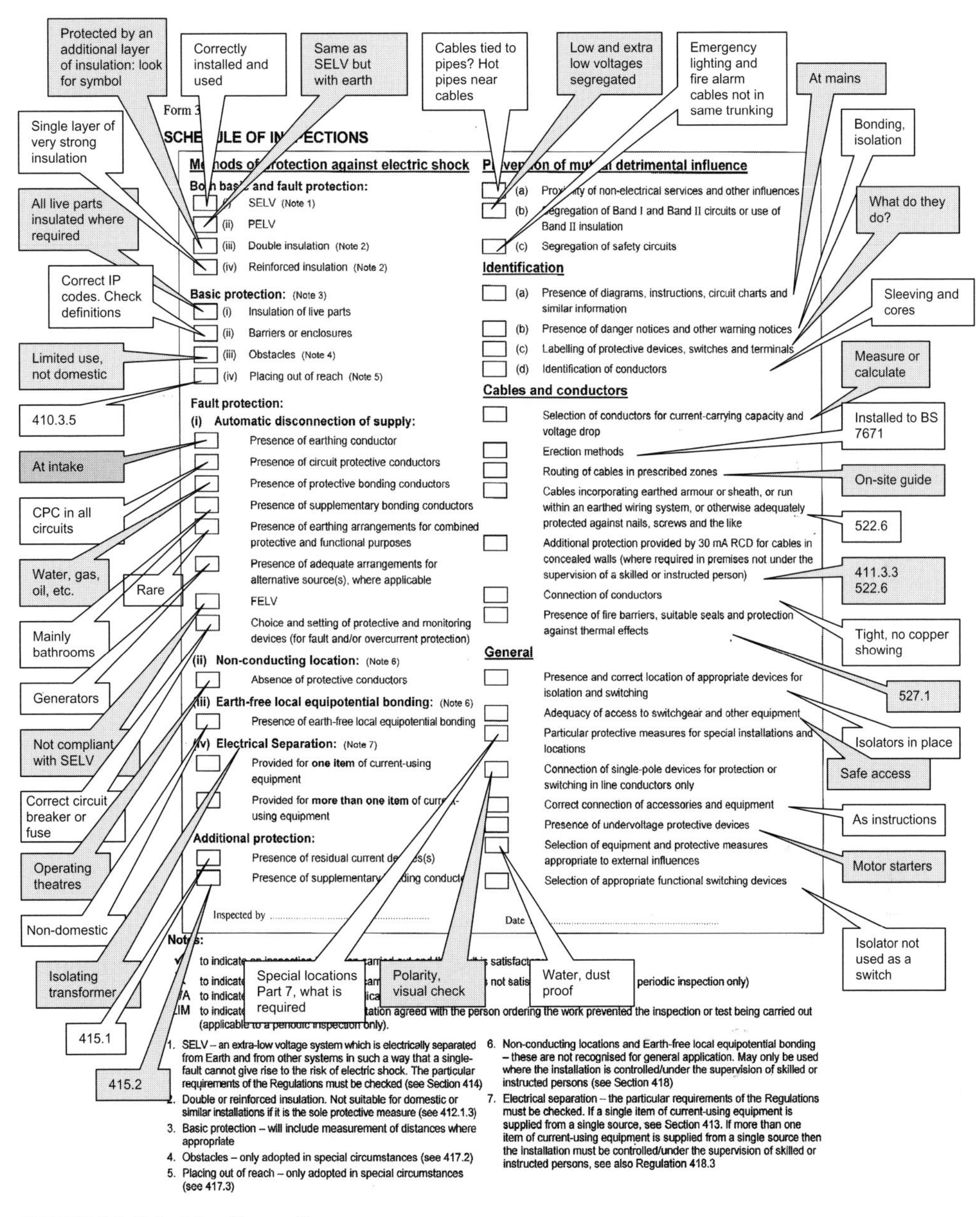

FIGURE 6.9 Schedule of inspections

Periodic inspection and testing

The purpose of this type of inspection is to ensure that the installation is safe for continued service and to identify any non-compliances.

A periodic inspection should be carried out on a regular basis, the date of the first periodic test being set by the designer of the installation. The interval between any subsequent tests should be set by the person carrying out the test. The inspection is carried out with minimum dismantling; the amount of dismantling will depend on the condition of the installation.

It is important that any past test results and certificates are available. Remember that the object of this inspection and test, apart from ensuring that it is safe, is to monitor its condition and identify any area of deterioration. This will not be possible unless we have the past test results to compare with our new results.

If there is no documentation available, then a survey must be carried out to complete any documentation that may be required, such as type and designation of circuits. We would also need to ensure that the installation is safe to work on. Of course, while the documentation is being prepared, an element of the inspection can be carried out at the same time.

Before commencing work on the inspection and testing, the amount of investigation which is going to be carried out must be agreed, along with any area which cannot be accessed or circuits which cannot be isolated. This must be recorded in the extent and limitations box which is part of the periodic inspection report (PIR). See Figure 6.10.

Once the extent and limitations have been agreed the inspection can begin and the areas of investigation must include:

- Safety
- Ageing
- Damage
- Corrosion
- Overload
- Wear and tear
- Additions
- Alterations.

The investigation may require an element of dismantling; this is fine but it should be kept to a minimum. Supplementary testing will also be required; this is another reason why past test results must be available, as, when tests are carried out we will need something to compare them to.

The first part of a visual inspection is to ensure that the system is safe to test and that you have enough information to be able to carry out the test safely.

Generally, a good place to start would be the supply intake; this will give a reasonable indication of the age, type and size of the installation.

Things to look for at the supply intake *before* removal of any covers would be:

- The type of supply system: is it TT, TN-S or TN-C-S?
- Is it old or modern?
- Are the conductors imperial or metric?
- What type of protection is there for the final circuits?
- Is documentation available for the original installation?
- Are the distribution boards labelled correctly?
- Is the earthing conductor in place?

IPM4

Original (To the person ordering the work)

PERIODIC INSPECTION REPORT FOR AN ELECTRICAL INSTALLATION

Issued in accordance with *British Standard BS 7671- Requirements for Electrical Installations*

A. DETAILS OF THE CLIENT

Client: Address:

B. PURPOSE OF THE REPORT

This Periodic Inspection Report must be used only for reporting on the condition of an existing installation.

Purpose for which this report is required:

C. DETAILS OF THE INSTALLATION

Occupier:

Address:

Postcode:

Description of premises: Domestic Commercial Industrial

Other: (Please state)

Estimated age of the electrical installation: years

Evidence of alterations or additions If yes, estimated age years

Date of previous inspection: Electrical Installation Certificate No or previous Periodic Inspection Report No:

Records of installation available: Records held by:

D. EXTENT OF THE INSTALLATION AND LIMITATIONS OF THE INSPECTION AND TESTING

Extent of the electrical installation covered by this report:

Agreed limitations (including the reasons), if any, on the inspection and testing:

This inspection has been carried out in accordance with BS 7671, as amended. Cables concealed within trunking and conduits, or cables and conduits concealed under floors, in inaccessible roof spaces and generally within the fabric of the building or underground, have not been visually inspected.

E. DECLARATION

I/We, being the person(s) responsible for the inspection and testing of the electrical installation (as indicated by my/our signatures below), particulars of which are described above (see C), having exercised reasonable skill and care when carrying out the inspection and testing, hereby declare that the information in this report, including the observations (see F) and the attached schedules (see H), provides an accurate assessment of the condition of the electrical installation taking into account the stated extent of the installation and the limitations of the inspection and testing (see D).

I/We further declare that in my/our judgement, the said installation was overall in ❖ condition (see G) at the time the inspection was carried out, and that it should be further inspected as recommended (see I).

❖ *(Insert* ***'a satisfactory'*** *or* ***'an unsatisfactory'****, as appropriate)*

INSPECTION, TESTING AND ASSESSMENT BY:

Signature:

Name: (CAPITALS)

Position:

Date:

REPORT REVIEWED AND CONFIRMED BY: † *See note below*

Signature:

Name: (CAPITALS)

Date:

† *The completed report should preferably be reviewed by another competent person to confirm that the declared overall condition of the electrical installation is consistent with the inspection and test results, and with the observations and recommendations for action (if any) made in the report.*

Page 1 of

This form is based on the model shown in Appendix 6 of BS 7671.

Please see the 'Notes for Recipients' on the reverse of this page.

IPM4/1

FIGURE 6.10 IPM4 PIR for an electrical installation

NOTES FOR RECIPIENTS

THIS SAFETY CERTIFICATE IS AN IMPORTANT AND VALUABLE DOCUMENT WHICH SHOULD BE RETAINED FOR FUTURE REFERENCE

The purpose of periodic inspection is to determine, so far as is reasonably practicable, whether an electrical installation is in a satisfactory condition for continued service. This report provides an assessment of the condition of the electrical installation identified overleaf at the time it was inspected, taking into account the stated extent of the installation and the limitations of the inspection and testing.

The report has been issued in accordance with the national standard for the safety of electrical installations, British Standard 7671 (as amended) - *Requirements for Electrical Installations*.

Where the installation incorporates a residual current device (RCD), there should be a notice at or near the main switchboard or consumer unit stating that the device should be tested at quarterly intervals. For safety reasons, it is important that you carry out the test regularly.

Also for safety reasons, the electrical installation will need to be re-inspected at appropriate intervals by a competent person. The recommended maximum time interval to the next inspection is stated on page 3 in Section I (*Next Inspection*). There should be a notice at or near the main switchboard or consumer unit indicating when the next inspection of the installation is due.

The report consists of at least six numbered pages. The report is invalid if any of the pages identified in Section H are missing.

For installations having more than one distribution board or more circuits than can be recorded on Pages 5 and 6, one or more additional *Schedules of Circuit Details for the Installation,* and *Schedules of Test Results for the Installation* (pages 7 and 8 onwards) should form part of the report.

This report is intended to be issued only for the purpose of reporting on the condition of an existing electrical installation. The report should identify, so far as is reasonably practicable and having regard to the extent and limitations recorded in Section D, any damage, deterioration, defects, dangerous conditions and any non-compliances with the requirements of the national standard for the safety of electrical installations which may give rise to danger. It should be noted that the greater the limitations applying to a report, the less its value.

The report should not have been issued to certify that a new electrical installation complies with the requirements of the national safety standard. An 'Electrical Installation Certificate' or a 'Domestic Electrical Installation Certificate'(where appropriate) should be issued for the certification of a new installation.

You should have received the report marked 'Original' and the electrical contractor should have retained the report marked 'Duplicate'.

If you were the person ordering the work, but not the user of the installation, you should pass this report, or a full copy of it including these notes, the schedules and additional pages (if any), immediately to the user.

The 'Original' report form should be retained in a safe place and shown to any person inspecting or undertaking further work on the electrical installation in the future. If you later vacate the property, this report will provide the new user with an assessment of the condition of the electrical installation at the time the periodic inspection was carried out.

Section D addresses the extent and limitations of the report by providing boxes for the *Extent of the electrical installation covered by this report* and the *Agreed limitations, if any, on the inspection and testing.* Information given here should fully identify the scope of the inspection and testing and of the report. The electrical contractor should have agreed all such aspects with the person ordering the work and other interested parties (eg licensing authority, insurance company, building society etc) before the inspection was carried out.

continued on the reverse of page 3

IPM4/1&2

FIGURE 6.10 (*Continued*)

IPM4

Original (To the person ordering the work)

F. OBSERVATIONS AND RECOMMENDATIONS FOR ACTIONS TO BE TAKEN

Referring to the attached schedules of inspection and test results, and subject to the limitations at D:

There are no items adversely affecting electrical safety. ☐

or

The following observations and recommendations are made. ☐

Item No		Code †
1		

Note: If necessary, continue on additional page(s), which must be identified by the Periodic Inspection Report date and page number(s).

† *Where observations are made, the inspector will have entered one of the following codes against each observation to indicate the action (if any) recommended:-*

1. *'requires urgent attention' or*
2. *'requires improvement' or*
3. *'requires further investigation' or*
4. *'does not comply with BS 7671'*

Please see the reverse of this page for guidance regarding the recommendations.

Urgent remedial work recommended for Items: ☐ **Corrective action(s) recommended for Items:** ☐

G. SUMMARY OF THE INSPECTION

General condition of the installation:

Note: If necessary, continue on additional page(s), which must be identified by the Periodic Inspection Report date and page number(s).

Date(s) of the inspection: ☐ Overall assessment of the installation: ☐

*(Entry should read either **'Satisfactory'** or **'Unsatisfactory'**)*

Page 2 of ☐

Please see the 'Guidance for Recipients on the Recommendation Codes' on the reverse of this page.

IPM4/3

FIGURE 6.10 (*Continued*)

GUIDANCE FOR RECIPIENTS ON THE RECOMMENDATION CODES

Only one Recommendation Code should have been given for each recorded observation.

Recommendation Code 1

Where an observation has been given a Recommendation Code 1 (requires urgent attention), the safety of those using the installation may be at risk.

The person responsible for the maintenance of the installation is advised to take action without delay to remedy the observed deficiency in the installation, or to take other appropriate action (such as switching off and isolating the affected part(s) of the installation) to remove the potential danger. The electrical contractor issuing this report will be able to provide further advice.

NICEIC make available 'dangerous condition' notification forms to enable inspectors to record, and then to communicate to the person ordering the report, any dangerous condition discovered.

Recommendation Code 2

Recommendation Code 2 (requires improvement) indicates that, whilst the safety of those using the installation may not be at immediate risk, remedial action should be taken as soon as possible to improve the safety of the installation to the level provided by the national standard for the safety of electrical installations, BS 7671. The electrical contractor issuing this report will be able to provide further advice.

Items which have been attributed Recommendation Code 2 should be remedied as soon as possible (see Section F).

Recommendation Code 3

Where an observation has been given a Recommendation Code 3 (requires further investigation), the inspection has revealed an apparent deficiency which could not, due to the extent or limitations of this inspection, be fully identified. Items which have been attributed Recommendation Code 3 should be investigated as soon as possible (see Section F).

The person responsible for the maintenance of the installation is advised to arrange for the further examination of the installation to determine the nature and extent of the apparent deficiency.

Recommendation Code 4

Recommendation Code 4 [does not comply with BS 7671 (as amended)] will have been given to observed non-compliance(s) with the **current** safety standard which do not warrant one of the other Recommendation Codes. It is not intended to imply that the electrical installation inspected is unsafe, but careful consideration should be given to the benefits of improving these aspects of the installation. The electrical contractor issuing this report will be able to provide further advice.

It is important to note that the recommendation given at Section I *Next Inspection* of this report for the maximum interval until the next inspection, is conditional upon all items which have been given a Recommendation Code 1 and Code 2 being remedied without delay, and as soon as possible respectively (see Section F).

It would not be reasonable to indicate a 'satisfactory' assessment if any observation in the report had been given a Code 1 or Code 2 recommendation (see Section G).

IPM4/3&4

FIGURE 6.10 (*Continued*)

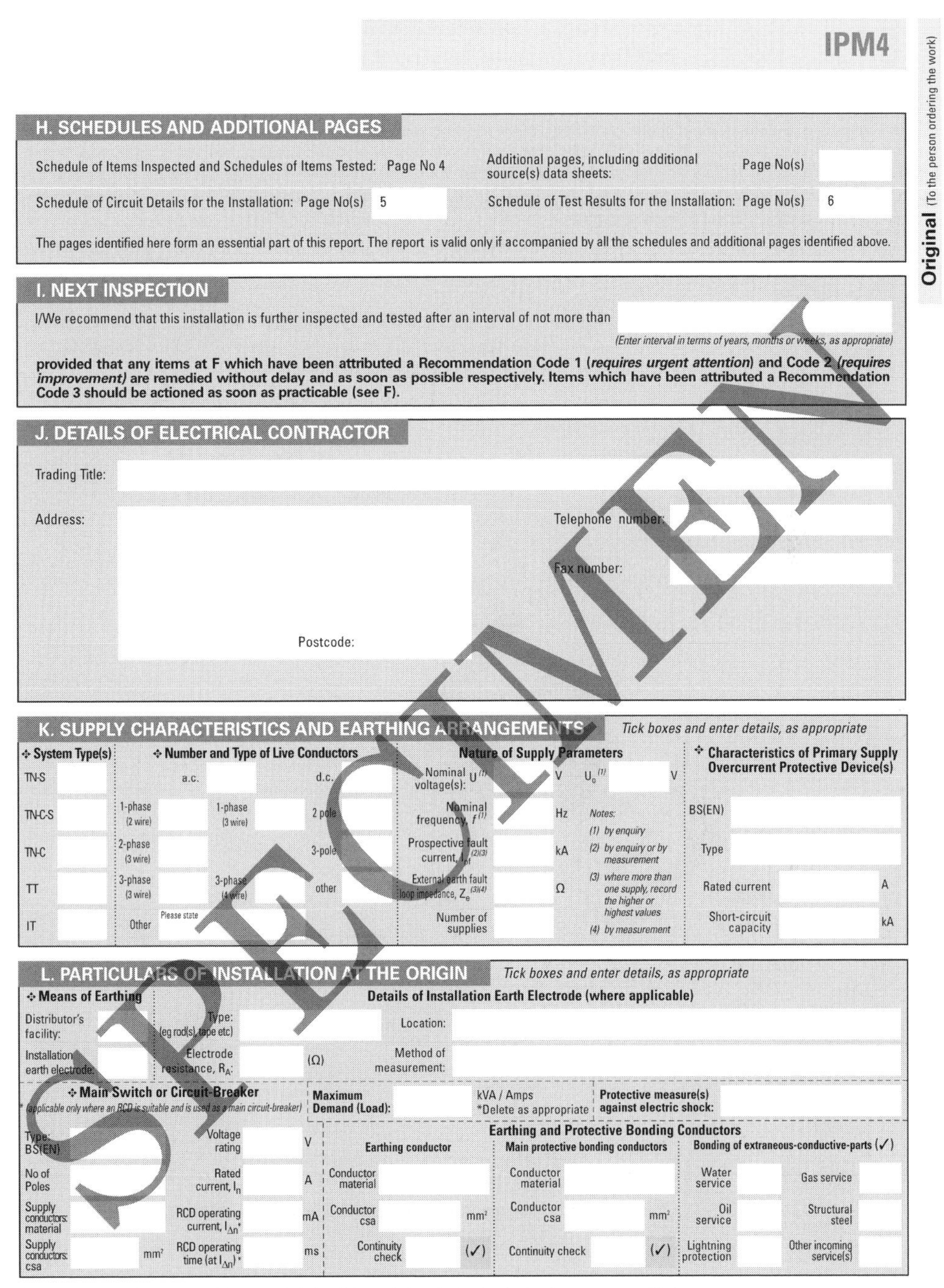

IPM4

Original (To the person ordering the work)

H. SCHEDULES AND ADDITIONAL PAGES

Schedule of Items Inspected and Schedules of Items Tested: Page No 4

Additional pages, including additional source(s) data sheets: Page No(s)

Schedule of Circuit Details for the Installation: Page No(s) 5

Schedule of Test Results for the Installation: Page No(s) 6

The pages identified here form an essential part of this report. The report is valid only if accompanied by all the schedules and additional pages identified above.

I. NEXT INSPECTION

I/We recommend that this installation is further inspected and tested after an interval of not more than

(Enter interval in terms of years, months or weeks, as appropriate)

provided that any items at F which have been attributed a Recommendation Code 1 (*requires urgent attention*) and Code 2 (*requires improvement*) are remedied without delay and as soon as possible respectively. Items which have been attributed a Recommendation Code 3 should be actioned as soon as practicable (see F).

J. DETAILS OF ELECTRICAL CONTRACTOR

Trading Title:

Address:

Telephone number:

Fax number:

Postcode:

K. SUPPLY CHARACTERISTICS AND EARTHING ARRANGEMENTS

Tick boxes and enter details, as appropriate

✣ System Type(s)	✣ Number and Type of Live Conductors			Nature of Supply Parameters		✣ Characteristics of Primary Supply Overcurrent Protective Device(s)
TN-S	a.c.		d.c.	Nominal voltage(s): U [1] V	U_0 [1] V	
TN-C-S	1-phase (2 wire)	1-phase (3 wire)	2 pole	Nominal frequency, f [1] Hz	Notes: (1) by enquiry	BS(EN)
TN-C	2-phase (3 wire)		3-pole	Prospective fault current, I_{pf} [2][3] kA	(2) by enquiry or by measurement	Type
TT	3-phase (3 wire)	3-phase (4 wire)	other	External earth fault loop impedance, Z_e [3][4] Ω	(3) where more than one supply, record the higher or highest values	Rated current A
IT	Other	Please state		Number of supplies	(4) by measurement	Short-circuit capacity kA

L. PARTICULARS OF INSTALLATION AT THE ORIGIN

Tick boxes and enter details, as appropriate

✣ Means of Earthing

Distributor's facility:

Installation earth electrode:

Details of Installation Earth Electrode (where applicable)

Type: (eg rod(s), tape etc)

Location:

Electrode resistance, R_A: (Ω)

Method of measurement:

✣ Main Switch or Circuit-Breaker

* (applicable only where an RCD is suitable and is used as a main circuit-breaker)

Maximum Demand (Load): kVA / Amps *Delete as appropriate

Protective measure(s) against electric shock:

Type: BS(EN)

No of Poles

Supply conductors: material

Supply conductors: csa mm²

Voltage rating V

Rated current, I_n A

RCD operating current, $I_{\Delta n}$* mA

RCD operating time (at $I_{\Delta n}$)* ms

Earthing and Protective Bonding Conductors

Earthing conductor		Main protective bonding conductors		Bonding of extraneous-conductive-parts (✓)	
Conductor material		Conductor material		Water service	Gas service
Conductor csa	mm²	Conductor csa	mm²	Oil service	Structural steel
Continuity check	(✓)	Continuity check	(✓)	Lightning protection	Other incoming service(s)

✣ *Where a number of sources are available to supply the installation, and where the data given for the primary source may differ from other sources, a separate sheet must be provided which identifies the relevant information relating to each additional source.*

Page 3 of

This form is based on the model shown in Appendix 6 of BS 7671.

Please see the 'Notes for Recipients' on the reverse of this page.

IPM4/5

FIGURE 6.10 (*Continued*)

NOTES FOR RECIPIENTS
(continued from the reverse of page 1)

A declaration of the overall condition of the installation should have been given by the inspector in Section E of the report. The declaration must reflect that given in Section G, which summarises the observations and recommendations made in Section F. A list of observations and recommendations for urgent remedial work and corrective action(s) necessary to maintain the installation in a safe working order should have been given in Section F, where appropriate. For further guidance on the recommendations, please see the reverse of page 2.

It remains the responsibility of the compiler of the report to ensure that the information provided on the report is factual, and that the declaration (in Section E) of the overall condition of the electrical installation to which the report relates is reasonable in all the circumstances.

Where the installation can be supplied by more than one source, such as the public supply and a standby generator, the number of supplies should have been recorded in the box entitled *Number of Supplies*, in Section K *Supply Characteristics and Earthing Arrangements* on page 3 of the report, and the *Schedule of Test Results* compiled accordingly.

IPM4/5&6

FIGURE 6.10 ***(Continued)***

IPM4

Original (To the person ordering the work)

SCHEDULE OF ITEMS INSPECTED

† See note below

PROTECTIVE MEASURES AGAINST ELECTRIC SHOCK

Basic and fault protection

Extra low voltage

- SELV
- PELV

Double or reinforced insulation

- Double or Reinforced Insulation

Basic protection

- Insulation of live parts
- Barriers or enclosures
- Obstacles **
- Placing out of reach **

Fault protection

Automatic disconnection of supply

- Presence of earthing conductor
- Presence of circuit protective conductors
- Presence of protective bonding conductors
- Presence of earthing arrangements for combined protective and functional purposes
- Presence of adequate arrangements for alternative source(s), where applicable
- FELV
- Choice and setting of protective and monitoring devices (for fault protection and/or overcurrent protection)

Non-conducting location **

- Absence of protective conductors

Earth-free equipotential bonding **

- Presence of earth-free equipotential bonding

Electrical separation

- For **one** item of current-using equipment
- For **more** than one item of current-using equipment **

Additional protection

- Presence of residual current device(s)
- Presence of supplementary bonding conductors

***** For use in controlled supervised/conditions only***

Prevention of mutual detrimental influence

- Proximity of non-electrical services and other influences
- Segregation of Band I and Band II circuits or Band II insulation used
- Segregation of Safety Circuits

Identification

- Presence of diagrams, instructions, circuit charts and similar information
- Presence of danger notices and other warning notices
- Labelling of protective devices, switches and terminals
- Identification of conductors

Cables and Conductors

- Selection of conductors for current carrying capacity and voltage drop
- Erection methods
- Routing of cables in prescribed zones
- Cables incorporating earthed armour or sheath or run in an earthed wiring system, or otherwise protected against nails, screws and the like
- Additional protection by 30mA RCD for cables concealed in walls (where required, in premises not under the supervision of skilled or instructed persons)
- Connection of conductors
- Presence of fire barriers, suitable seals and protection against thermal effects

General

- Presence and correct location of appropriate devices for isolation and switching
- Adequacy of access to switchgear and other equipment
- Particular protective measures for special installations and locations
- Connection of single-pole devices for protection or switching in line conductors only
- Correct connection of accessories and equipment
- Presence of undervoltage protective devices
- Selection of equipment and protective measures appropriate to external influences
- Selection of appropriate functional switching devices

SCHEDULE OF ITEMS TESTED

† See note below

- External earth fault loop impedance, Z_e
- Installation earth electrode resistance, R_A
- Continuity of protective conductors
- Continuity of ring final circuit conductors
- Insulation resistance between live conductors
- Insulation resistance between live conductors and Earth
- Protection by SELV, PELV or by electrical separation
- Basic protection by barrier or enclosure provided during erection
- Insulation of non-conducting floors or walls
- Polarity
- Earth fault loop impedance, Z_s
- Verification of phase sequence
- Operation of residual current devices
- Functional testing of assemblies
- Verification of voltage drop

† ***All boxes must be completed.***

*'✔' indicates that an inspection or a test was carried out and that the result was **satisfactory***

*'✘' indicates that an inspection or a test was carried out and that the result was **unsatisfactory***

*'N/A' indicates that an inspection or a test was **not applicable** to the particular installation*

*'LIM' indicates that, that exceptionally, a **limitation** agreed with the person ordering the work (as recorded in Section D) **prevented** the inspection or test being carried out.*

Page 4 of

This form is based on the model shown in Appendix 6 of BS 7671

IPM4/7

FIGURE 6.10 *(Continued)*

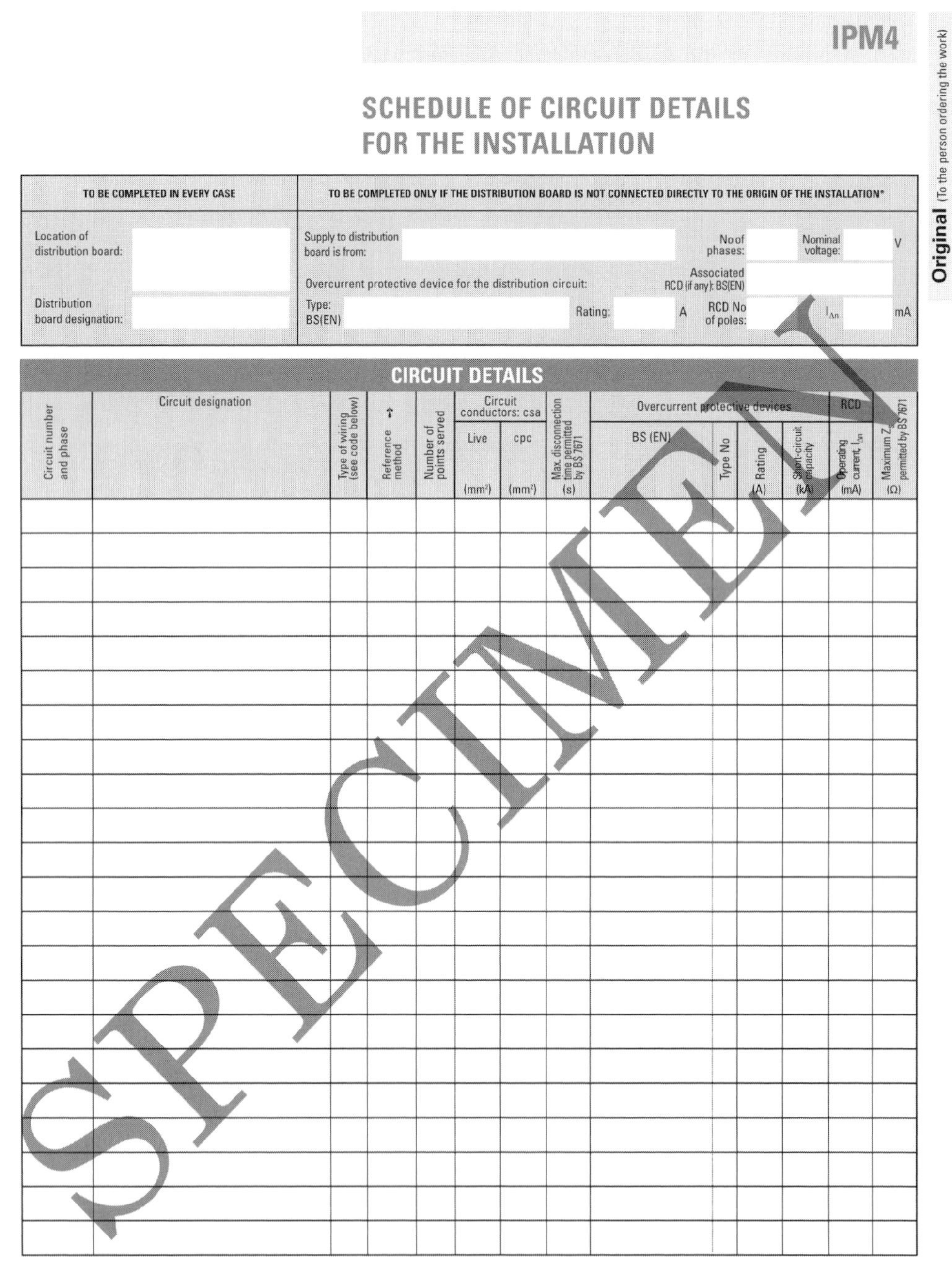

IPM4

Original (To the person ordering the work)

SCHEDULE OF CIRCUIT DETAILS FOR THE INSTALLATION

TO BE COMPLETED IN EVERY CASE	TO BE COMPLETED ONLY IF THE DISTRIBUTION BOARD IS NOT CONNECTED DIRECTLY TO THE ORIGIN OF THE INSTALLATION*
Location of distribution board:	Supply to distribution board is from: No of phases: Nominal voltage: V
	Overcurrent protective device for the distribution circuit: Associated RCD (if any): BS(EN)
Distribution board designation:	Type: BS(EN) Rating: A RCD No of poles: $I_{\Delta n}$ mA

CIRCUIT DETAILS

Circuit number and phase	Circuit designation	Type of wiring (see code below)	Reference method †	Number of points served	Circuit conductors: csa Live (mm²)	Circuit conductors: csa cpc (mm²)	Max. disconnection time permitted by BS 7671 (s)	Overcurrent protective devices BS (EN)	Type No	Rating (A)	Short-circuit capacity (kA)	RCD Operating current, $I_{\Delta n}$ (mA)	Maximum Z_s permitted by BS 7671 (Ω)

† *See Table 4A2 of Appendix 4 of BS 7671.*

CODES FOR TYPE OF WIRING

A	B	C	D	E	F	G	H	O (Other - please state)
PVC/PVC cables	PVC cables in metallic conduit	PVC cables in non-metallic conduit	PVC cables in metallic trunking	PVC cables in non-metallic trunking	PVC/SWA cables	XLPE/SWA cables	Mineral-insulated cables	

Page 5 of

** In such cases, details of the distribution (sub-main) circuit(s), together with the test results for the circuit(s), must also be provided.*

This form is based on the model shown in Appendix 6 of BS 7671.

See next page for Schedule of Test Results

IPM4/9

FIGURE 6.10 ***(Continued)***

IPM4

Original (To the person ordering the work)

SCHEDULE OF TEST RESULTS FOR THE INSTALLATION

TO BE COMPLETED ONLY IF THE DISTRIBUTION BOARD IS NOT CONNECTED DIRECTLY TO THE ORIGIN OF THE INSTALLATION				
Characteristics at this distribution board				
	Confirmation of supply polarity			
* See note below				
Z_s *	Ω	Operating times of associated RCD (if any)	At $I_{\Delta n}$	ms
I_{pf} *	kA		At $5I_{\Delta n}$ (if applicable)	ms

Test instruments (serial numbers) used:			
Earth fault loop impedance		RCD	
Insulation resistance		Other	
Continuity		Other	

TEST RESULTS

Circuit number and phase	Circuit impedances (Ω)					Insulation resistance † Record lower or lowest value				Polarity	Maximum measured earth fault loop impedance, Z_s * See note below	RCD operating times	
	Ring final circuits only (measured end to end)			All circuits (At least one column to be completed)		Line/Line †	Line/Neutral †	Line/Earth †	Neutral/Earth			at $I_{\Delta n}$	at $5I_{\Delta n}$ (if applicable)
	r_1 (Line)	r_n (Neutral)	r_2 (cpc)	$R_1 + R_2$	R_2	(MΩ)	(MΩ)	(MΩ)	(MΩ)	(✓)	(Ω)	(ms)	(ms)

* *Note: Where the installation can be supplied by more than one source, such as a primary source (eg public supply) and a secondary source (eg standby generator), the higher or highest values must be recorded.*

TESTED BY

Signature:		Position:	
Name: (CAPITALS)		Date of testing:	

Page 6 of

This form is based on the model shown in Appendix 6 of BS 7671.

See previous page for Schedule of Circuit Details

IPM4/11

FIGURE 6.10 ***(Continued)***

- What size is the earthing conductor?
- Is the earthing conductor green, or green and yellow?
- Are all of the circuits in one consumer unit or are there two or three units that need combining?
- Is there any evidence of equipotential bonding? Remember: it must start at the main earthing terminal.
- What size is the equipotential bonding? Is it large enough?
- Is there an RCD? If so, has it a label attached? Is it a voltage or current operated type?
- Do the enclosures meet required IP codes? (Regulation 412.2.2.3.)
- If alterations have been carried out is there documentation available for them, along with test results?
- Where alterations have been carried out since January 2005, has a warning notice been fitted on or near to the distribution board to indicate that new colours have been used? (Regulation 514.14.1.)
- What size is the supply fuse? Is it large enough for the required load?
- Are the meter tails large enough?
- Are the seals broken on supply equipment? If they are it could indicate that the system has been tampered with since it was first installed and perhaps closer investigation is required.
- Have any alterations or additions been made?
- Would any alterations or additions affect the required disconnection time for the circuit concerned?

This list is not exhaustive and installation conditions may require more.

When the visual inspection of the supply intake area is complete, it is a good time to look around the building to make sure that there are no very obvious faults.

All of this should be carried out before removal of any covers.

Areas of inspection:

- Are any accessories missing or damaged? Are they fixed to the wall properly?
- Are the accessories old with wooden back plates?
- Are the socket outlets round pin or square? Is there a mixture of both?
- Have cables been installed in vulnerable situations?
- Have cables, enclosures and accessories been fixed securely?
- Have ceiling roses got perished flexes? Particular attention should be given to the old braided and rubber type flexes.
- Are all of the socket outlets RCD protected? If they were installed before 2008, it is not a requirement that they are, but an RCD should be listed as a recommendation unless there is a good reason for them not to be protected.
- Earthing clamps, where used, must be to BS 951 and correctly labelled.
- If gas or water is bonded using the same conductor, ensure that the conductor is continuous and not cut at the clamp.
- Is the supplementary bonding in place in the bathroom? Is it required? See regulation 701.415.2.

- Correct equipment for zones in bathroom/shower room (see 701 BS 7671).
- Are socket outlets installed in a room containing a bath or shower? If so, are the socket outlets 3 m from zone 1 and RCD protected?
- Is there any evidence of mutual detrimental influence? Are there any cables fixed to water, gas or any other non-electrical services? (*The cables need to be far enough away to avoid damage if the non-electrical services are worked on.*)
- Are cables of different voltage bands segregated? Low voltage, SELV, telephone cables or television aerials should not be fixed together. *(They are permitted to cross.)*

Whilst these items are being checked look in any cupboards for sockets or lights. If your customer is uncomfortable with this it is vitally important that you document any area that cannot be investigated in the extent and limitations section on the PIR.

During this purely visual part of the inspection you will gain some idea of the condition of the installation and identify any alterations or additions.

This list is not exhaustive but will not require removal of any fittings, etc.

Once you are happy that the installation is safe to work on, a more detailed visual inspection can be carried out and the dreaded but necessary form filling can be started.

Once again begin at the consumer unit.

Before you start it must be isolated: the EAWR 1989 states that it is an offence to work live. Once you remove a cover you will be working live if you do not isolate it.

Having carried out the safe isolation procedure, remove the cover of the consumer unit.

- Your first impression will be important: has care been taken over the terminations of cables (*neat and not too much exposed conductor*)?
- Are all cables terminated (*no loose ends*) and all connections tight?
- Are there any signs of overheating?
- Is there a mixture of protective devices?
- Are there any rubber cables?
- Are there any damaged cables (*perished or cut*)?
- Have all circuits got CPCs?
- Are all earthing conductors sleeved?
- On a photocopy of a schedule of test results record circuits, protective devices and cable sizes.
- Look to see if the protective devices seem suitable for the cable sizes that they are protecting.
- Note any type D or 4 miniature circuit breakers; these will require further investigation.
- Are all barriers in place?
- Have all of the circuit conductors been connected in sequence, with phase, neutral and CPC from circuit number 1 being in terminal number 1, preferably the highest current nearest the main switch?
- Have any protective devices got multiple conductors in them? Are they the correct size (*all the same*)?

- Is there only one set of tails or has another board been connected to the original board by joining at the terminals?

Having had a detailed look at the consumer unit, **with the installation still isolated** carry out a more detailed investigation of the rest of the installation.

It may be that you have agreed with your client that only 10% of the installation is to be inspected. This would mean 10% of each circuit. There would be little point in inspecting 10% of the circuits. If the period between inspections was 10 years, it could be many years before a circuit was eventually inspected and the exercise would be pointless.

During your preliminary walk around, you will have identified any areas of immediate concern, and these will be addressed as your inspection progresses. There is no reason why you should not start any dead testing that is required, as you progress through your visual inspection.

On radial circuits this would be a good time to carry out CPC continuity, $R_1 + R_2$, insulation resistance and polarity tests as you work your way round.

Start at circuit number 1 and work your way through the circuits one at a time.

But first, what are you looking for? Let us look at a selection of circuits.

Shower circuit

- Is isolation provided? If so, is it within prescribed zones?
- Has the correct size cable/protective device been selected?
- Is it bonded?
- Are connections tight?
- Has earth sleeving been fitted?
- Is the shower secure?
- Is there any evidence of water ingress?
- Is the shower in a bedroom?

Cooker circuit

- Is the switch within 2 m of the cooker or hob?
- Has the cooker switch got a socket outlet? If so, it requires a 0.4-s disconnection time.
- Green and yellow sleeving.
- If it has a metal faceplate, has it got an earthing tail to the flush box?
- Is the cable the correct size for a protective device?
- Are there any signs of overheating around the terminations?
- Is the cooker outlet too close to the sink? Building regulations require any outlets installed after January 2005 to be at least 300 mm from the sink.

Socket outlets

- Is there correct coordination between protective devices and conductors?
- Green and yellow sleeving fitted.

- Any metal sockets have an earthing tail back to the socket box.
- Radial circuit not serving too large an area (*see table 8A of the on-site guide*).
- Secure connections.
- Cables throughout the circuit are the same size.
- Are there any sockets outside? Are they waterproof? Are they 30-mA RCD protected?
- Are there any outlets in the bathroom? Are they SELV or at least 3 m from Zone 1 and RCD protected?
- Are there socket outlets within 3 m of a shower installed in a bedroom? If so, are they 30-mA RCD protected?
- Will the protective device for the circuit provide 0.4-s disconnection time?

Immersion heater circuits

- Is there correct coordination between the protective device and live conductors?
- Has the CPC been sleeved?
- Has the immersion heater got a thermal cut-out?
- Is the immersion the only equipment connected to this circuit? (*Any water heater with a capacity of 15 litres or more must have its own circuit; on-site guide appendix 8*). Often you will find that the central heating controls are supplied through the immersion heater circuit, which is incorrect.
- Is the immersion heater connected with heat-resistant cord?
- The immersion heater switch should be a cord outlet type, not a socket outlet and plug.
- If the supplementary bonding for the bathroom is carried out in the cylinder cupboard does the supplementary bonding include the immersion heater switch? It should.

Lighting circuits

- Is there correct coordination between the protective device and the live conductors?
- How many points are there on the circuit? A rating of 100 W minimum must be allowed for each lighting outlet. Shaver points, clock points and bell transformers may be neglected for the purpose of load calculation. As a general rule, 10 outlets per circuit is about right. Also remember that fluorescent fittings and discharge lamps are rated by their output, and the output must be multiplied by a factor of 1.8 if exact information is not available (*table 1A of the on-site guide*).
- Are the switch returns colour identified at both ends?
- Have the switch drops got CPCs? If they have, are they sleeved with green and yellow?
- Are the CPCs correctly terminated?
- Are the switch boxes made of wood or metal?
- Ceiling roses must be suitable for the mass hanging from them.

- Only one flexible cord should come out of each ceiling rose unless it is designed for multiple cords.
- Light fittings in bathrooms must be suitable for the zones in which they are fitted.
- Is the phase conductor to ES lampholders connected to the centre pin? This does not apply to E14 and E27 lampholders (*regulation 612.6*).

Three-phase circuits/systems

These circuits should be inspected for the same defects that you could find in other circuits. In addition to this:

- Are warning labels fitted where the voltage will be higher than expected? For example, a lighting switch with two phases in it, or perhaps where sockets close to each other are on different phases.
- Are conductors in the correct sequence?
- Remember PFC should be double the phase to neutral fault current.

Occasionally, other types of circuit will be found, but the same type of inspection should be carried out using common sense.

Always remember that the reason for this inspection is to ensure safety.

Periodic testing

The level of testing will usually be far less for periodic testing than for initial verification; this is provided that previous test results are available. If they are not, then it will be necessary for the full survey and the complete range of tests to be carried out on the installation, to provide a comprehensive set of results.

The level of testing will depend largely on what the inspector discovers during the visual inspection, and the value of results obtained while carrying out sample testing. If any tests show significantly different results, then further testing may be required.

In some cases all of the installation will need to be tested.

Periodic testing can cause danger and due consideration should be given to safety. Persons carrying out the testing must be competent and experienced in the type of installation being tested and the test instruments being used.

There is no set sequence for the testing which may be required for the completion of the PIR. The sequence and the type of tests to be carried out are left to the person carrying out the test to decide upon.

Where tests are required the recommendations for these tests are as shown in Table 6.3.

Where there are no past results comparison will not be possible; an example of this would be carrying out an earth fault loop impedance test on a ring circuit. Where the results of the test are identical to the results on the existing certificate, no further testing of the ring circuit is required provided we are sure that our instruments are accurate. If we are obtaining the same results, what could have changed? Why would we need to carry out a ring final circuit test? Common sense tells me that if the visual inspection did not show any defects and the Z_s values for the circuit have not changed

TABLE 6.3 Recommendations for periodic testing

Inspection	Recommended tests
Continuity of protective conductors	Between the distribution board earth terminal and exposed conductive parts of current-using equipment
	Earth terminals of socket outlets (test to the fixing screw of outlet for convenience)
Continuity of bonding conductors	All main bonding and supplementary bonding conductors
Ring circuit continuity	Only required where alterations or additions have been made to the ring circuit
Insulation resistance	Only between live conductors joined and earth
	Or between live conductors with the functional switch open if testing lighting circuit
Polarity	Live polarity tested at the origin of the installation
	Socket outlets
	At the end of radial circuits
	Distribution boards
Earth electrode resistance	Isolate installation and remove earthing conductor to avoid parallel paths
Earth fault loop impedance	At the origin of the installation for Z_e
	Distribution boards for the Z_e of that board
	Socket outlets and at the end of radial circuits for Z_s
Functional tests	RCD tests and manual operation of isolators, protective devices and switches

dramatically, all that is required is to record the new Z_s value and then copy across the $R_1 + R_2$ values from the old certificate to the new.

This procedure is the same for all circuits; remember we are monitoring the safety of the installation, and there is no value to be gained by taking it all apart. We may not put it back together correctly as mistakes do happen, and we may also cause damage to the installation.

Where testing is required, there is no set sequence. The requirement is that it is carried out safely.

On completion of the inspection, a PIR must be completed indicating the condition of the installation along with any observations and recommendations.

Observations are categorised from 1 to 4 as follows:

1. Requires urgent attention
2. Requires improvement
3. Requires further investigation

4. Does not comply with the current edition of BS 7671; but this does not make the installation unsafe.

Very often the category which is given to an observation is a matter of making a sensible judgement, which will depend on the type of use and environmental conditions, as well as whether the person using the installation is a skilled or an ordinary person. It is vital for the inspector to have suitable knowledge and experience in the type of installation which is being inspected to enable the correct observation to be made.

A general list of codes is provided here. These categories should not be taken as being cast in stone, and the final decision is always down to the inspector; it is often a point of discussion.

Code 1 (requires urgent attention)

Code 1 is used to indicate that danger exists, requiring urgent action to remove it.

Where the persons using the installation are at risk, the person ordering the report should be advised to take action without delay to remedy the observed danger in the installation or to take other appropriate action (such as switching off and isolating the affected parts of the installation) to remove the danger.

The person carrying out the inspection should report any dangerous situations immediately to the person ordering the report. Do not wait until the report is completed.

Code 1 observations include:

- Exposed live parts that are accessible to touch, such as where a fuse carrier or circuit breaker is omitted from a consumer unit and a blanking piece is not fitted in its place; this condition would not satisfy the requirements of regulation 416.2.1.
- Terminations or connections have no barriers or enclosures, such as those belonging to a consumer unit.
- Live conductors have damaged insulation.
- An accessory is damaged.
- Absence of an effective means of earthing for the installation.
- An RCD fails to operate when the integral test button is operated.
- Excessive heat from electrical equipment causing damage to the installation or its surroundings.
- Incorrect polarity or a protective device in the neutral conductor only.
- Circuits with incorrect overcurrent protection (due, for example, to oversized fuse wire in rewirable fuses).
- Absence of RCD protection for socket outlets in rooms containing a bath or shower, other than SELV or shaver socket outlets (periodic only, RCD required in these rooms for all new installations).
- Socket outlets other than SELV or shaver socket outlets located within 3 m horizontally from the boundary of zone 1 in a location containing a bath or shower.

Code 2 (requires improvement)

This code is used to indicate that the observed non-compliance requires action to remove potential danger.

The person ordering the report should be advised that, whilst the safety of those using the installation may not be immediately at risk, remedial action should be taken as soon as possible to ensure the safety of the installation. Code 2 (requires improvement) recommendations include:

- A 30/32-A ring final circuit discontinuous or cross-connected with another circuit.
- Separate protective devices in line and neutral conductors (for example, double pole fusing).
- A public utility water pipe being used as the means of earthing for the installation.
- A gas or oil pipe being used as the means of earthing for the installation.
- Absence of a CPC for a lighting circuit supplying one or more items of Class I equipment[1].
- Size of earthing conductor does not satisfy adiabatic requirements (that is, does not comply with regulation 543.1.1).
- Absence of a CPC for a circuit, other than a lighting circuit, supplying one or more items of Class I equipment.
- Absence of earthing at a socket outlet.
- Absence of main protective bonding (except to a lightning protection system conductor, where a Code 3 recommendation may be appropriate).
- Absence of fault protection (protection against indirect contact) by an RCD where required, such as for a socket outlet circuit in an installation or cables buried in walls less than 50 mm deep and not protected by earthed metal.
- A 'borrowed neutral', for example where a single final circuit neutral is shared by two final circuits (such as an upstairs lighting circuit and a separately protected downstairs lighting circuit).
- Absence of a warning notice indicating the presence of a second source of electricity, such as a micro-generator.
- Fire risk from incorrectly installed electrical equipment, including incorrectly installed downlighters.
- Undersized main protective bonding conductors, where the conductor is less than 6 mm^2 or where there is evidence of thermal damage.
- Unenclosed connections at luminaires. (Such a defect can contribute to a fire, particularly where extra-low-voltage filament lamps are used.)
- Immersion heater does not comply with BS EN 60335-2-73 (that is, it does not have a built in cut-out that will operate if the stored water temperature reaches 98°C when the thermostat fails), and the cold water storage tank is plastic.
- Unsatisfactory functional operation of equipment where this may result in danger.
- Absence of RCD protection for portable or mobile equipment that may reasonably be expected to be used outdoors.

---

1. Where there is no CPC on a lighting circuit which only has Class II equipment connected it would be a Code 4 observation.

- Earth fault loop impedance value greater than that required for operation of the protective device within the time prescribed in the version of BS 7671/IEE Wiring Regulations current at the time of installation (use of Z_s tables from 16th edition required).
- Insulation of live conductors deteriorated to such an extent that the insulating material readily breaks away from the conductors (perished insulation).
- Absence of supplementary bonding where required, such as in a bathroom or shower room, where **all** the following three conditions are **not** satisfied:
  - All final circuits of the location comply with the requirements of regulation 411.3.2 for automatic disconnection.
  - All final circuits of the location have additional protection by means of a 30-mA RCD.
  - All extraneous conductive parts of the location are effectively connected to the protective equipotential bonding (main earthing terminal).

Note: It is not always possible to identify that supplementary bonding is in place visually; a continuity test is often the only option.

Code 3 (requires further investigation)

A Code 3 could be used to indicate that the inspector was unable to come to a conclusion about an aspect of the installation or, alternatively, that the observation was outside the agreed purpose, extent or limitations of the inspection, but has come to the inspector's attention during the inspection and testing.

The person ordering the report should be advised that the inspection has revealed an apparent deficiency which could not, due to the agreed extent or limitations of the inspection, be fully identified, and that the deficiency should be investigated as soon as possible.

A Code 3 recommendation would usually be associated with an observation on a part of the installation that was not foreseen when the extent and limitations of the inspection were agreed with the client.

The purpose of a PIR is to assess the safety of the installation for ongoing use, within the extent and limitations agreed with the person ordering the work. On occasions, due to the condition of the installation, the inspector may advise the client that further inspection is required.

Observations that would usually warrant a Code 3 recommendation include:

- Unable to trace final circuits.
- Unable to access equipment or connections needing to be inspected that are known to exist but have been boxed in such as by panels or boards that cannot be easily removed without causing damage to decorations.
- Insulation resistance of less than 1 MΩ between live conductors connected together and earth, when measured at the consumer unit with all final circuits connected.
- Absence of a main protective bonding connection to a lightning protection system conductor, where it is not known by the inspector if it is required to protect against

lightning side flashes[2]. (Absence of other main equipotential bonding connections would usually warrant a recommendation Code 2.)

Code 4 (does not comply with the current issue of BS 7671)

This code is used to where certain items have been identified as not complying with the requirements of the current issue of BS 7671, but the users of the installation are not in any danger as a result.

Advice should be given to the person ordering the work with regard to any Code 4 observation with an explanation as to why it does not comply, and what work could be carried out to improve the installation.

Observations that would usually warrant a Code 4 recommendation include:

- Switch lines not identified as line conductors at terminations (for example, a conductor having blue insulation is not sleeved brown in switches or lighting points).
- CPCs or final circuit conductors in a consumer unit not arranged or marked so that they can be identified for inspection, testing or alteration of the installation.
- Undersized main protective bonding conductors (subject to a minimum size of $6\,mm^2$), if there is no evidence of thermal damage.
- Absence of CPCs in circuits having only Class II (or all insulated) luminaires and switches.[1]
- Protective conductor of a lighting circuit not (or incorrectly) terminated at the final circuit connection point to a Class II (or insulated) item of equipment, such as at a switch mounting box or luminaire.
- Absence of 'Safety Electrical Connection – Do not remove.'
- Absence of a notice indicating that the installation has wiring colours to two versions of BS 7671.
- Absence of RCD periodic test notice.
- Absence of circuit identification details.
- Sheath of an insulated and sheathed non-armoured cable not taken inside the enclosure of an accessory, such as at a socket outlet or lighting switch. (Note: A Code 2 recommendation would be warranted if unsheathed cores are accessible to touch and/or likely to come into contact with metalwork.)
- Bare protective conductor of an insulated and sheathed cable not sleeved with insulation, colour coded to indicate its function.
- Installation not divided into an adequate number of circuits to minimise inconvenience for safe operation, fault clearance, inspection, testing and maintenance.
- Inadequate number of socket outlets. (Code 2 if extension leads run through a doorway, wall or window.)

---

2. Unless you are a specialist in lightning conductors, it is always better to seek advice as to whether it should or should not be bonded.

- Fixed equipment does not have a means of switching off for mechanical maintenance, where such maintenance involves a risk of burns or injury from mechanical movement.
- Absence of supplementary bonding to the installed Class II equipment where required (such as in a bathroom or shower room), in case the equipment is replaced with Class I equipment in the future.
- Reliance on a voltage-operated earth leakage circuit breaker for fault protection (protection against indirect contact), subject to the device being proved to operate correctly. (If the circuit breaker relies on a water pipe not permitted by regulation 542.2.4 as the means of earthing, this would attract a Code 2 recommendation.)
- Absence of RCD protection for cables installed at a depth of less than 50 mm from a surface of a wall or partition where the cables do not incorporate an earthed metallic covering, are not enclosed in earthed metalwork or are not mechanically protected against penetration by nails and the like.
- Absence of RCD protection for cables concealed, at whatever depth, in a wall or partition the internal construction of which includes metallic parts (other than metallic fixings such as nails, screws and the like) where the cables do not incorporate an earthed metallic covering, are not enclosed in earthed metalwork or are not mechanically protected to avoid damage to them during construction of the wall or during their installation.
- Absence of RCD protection for socket outlet circuits that are unlikely to supply portable or mobile equipment for use outdoors or that are in a bathroom or shower room.
- Main protective bonding to gas, water or other service pipe is inaccessible for inspection, testing and maintenance, or connection not made before any branch pipework. (Note: The connection should preferably be within 600 mm of the meter outlet union or at the point of entry to the building if the meter is external.)
- Use of unsheathed flex for lighting pendants.
- Socket outlet mounted such as to result in potential damage to socket, plug and/or flex.
- No earth tail link between the earthing terminal of an insulated accessory and a metal back box, provided there is one fixed lug and the CPC is not formed by conduit, trunking, ducting or the metal sheath and/or armour of a cable.
- Installation divided into too few circuits.

As well as the PIR (Figure 6.8), a schedule of inspection and test results must be completed. Most certification bodies include the schedule of inspection and tests as part of the PIR. BS 7671 provides separate documents for each in appendix 6.

Chapter 7

Part 7

In this Chapter:

SPECIAL INSTALLATIONS OR LOCATIONS

These are installations or locations where due to the nature of the environment there is a greater risk of fire or electric shock.

Part 7 of BS 7671 contains 15 sections with information on installations or locations which are subject to additional considerations. Within these locations, additional work is required which must be carried out to provide additional protection. It is important to remember that the requirements of these special locations are to be provided to supplement or modify the requirements of BS 7671 parts 1 and 6 and are not to be used instead of them.

The numbering system used in part 7 can generally be cross-referenced with the requirements of the general regulations. For example, section 709.512.2, containing information regarding external influences which should be considered for marinas, should be used along with section 512.2 which provides information on the general requirements for external influences. This will ensure complete compliance with BS 7671.

SECTION 701

Locations containing a bath or shower

Bathrooms and shower rooms are considered a special location because apart from the kitchen, these areas may contain more extraneous conductive parts than other

A Practical Guide to the 17th Edition of the Wiring Regulations. DOI: 10.1016/B978-0-08-096560-4.00007-2

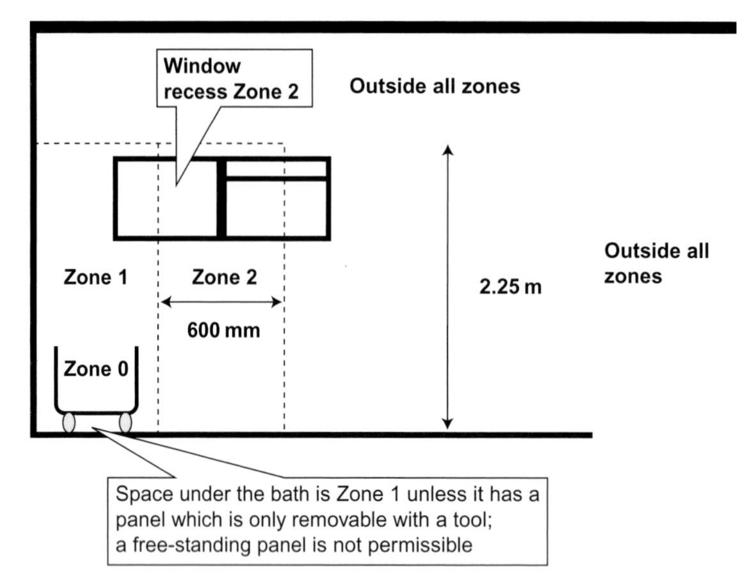

FIGURE 7.1 **Bathroom zones**

areas, and will also be wet or damp. We can add to this the fact that while using the bathroom people are generally only partially clothed or even naked. Wet skin and metalwork under fault conditions are clearly very dangerous; this is why additional precautions are required.

Bathrooms and shower rooms are divided into Zones 0, 1 and 2. The lower the number the more onerous the conditions are deemed to be.

Zone 0 is actually within the bath or shower tray. In a shower room without a tray, Zone 0 will be an area up to 100 mm from the floor.

Zone 1 is the area immediately above the bath or shower up to a height of 2.25 m from floor level or to the height of the shower head, whichever is the higher. However, personally I would feel uncomfortable installing any Class I equipment above Zone 1 which could be touched by a person standing in the bath.

Where a shower room does not have a tray and relies on a tiled floor only, Zone 1 is extended horizontally to a distance of 1.2 m from the centre point of the water outlet.

Zone 2 is the area up to 600 mm horizontally from Zone 1, up to a height of 2.25 m. The space above 2.25 m or further away from Zone 1 than 600 mm is deemed to be outside all zones.

We have to be careful because areas within these rooms which are not given zones will still require special consideration to ensure that the environment is as safe as we can possibly make it without restricting its use unreasonably (Figures 7.1–7.3).

The first thing that we need to remember is that we must ensure that the circuits in the bathrooms comply with the requirements for automatic disconnection as listed in part 4 of the regulations, unless the protective measure used is extra low voltage provided by SELV or PELV (regulation 701.414). In bathrooms, this will be a maximum voltage of 12 V a.c. or 30 V d.c.

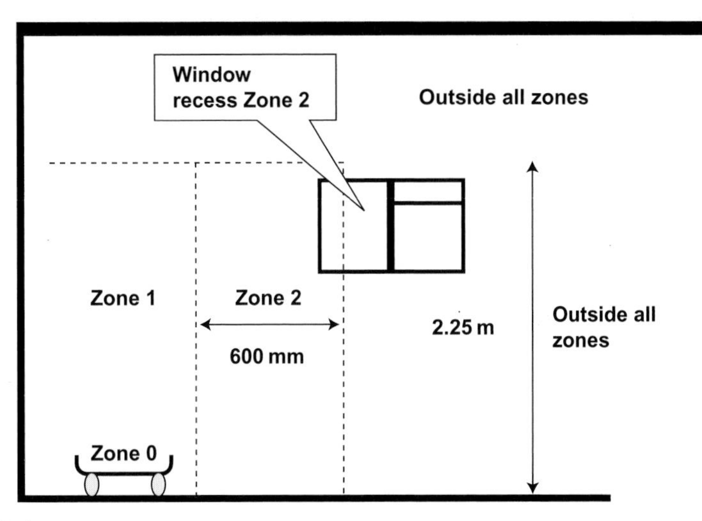

FIGURE 7.2 **Bathroom zones**

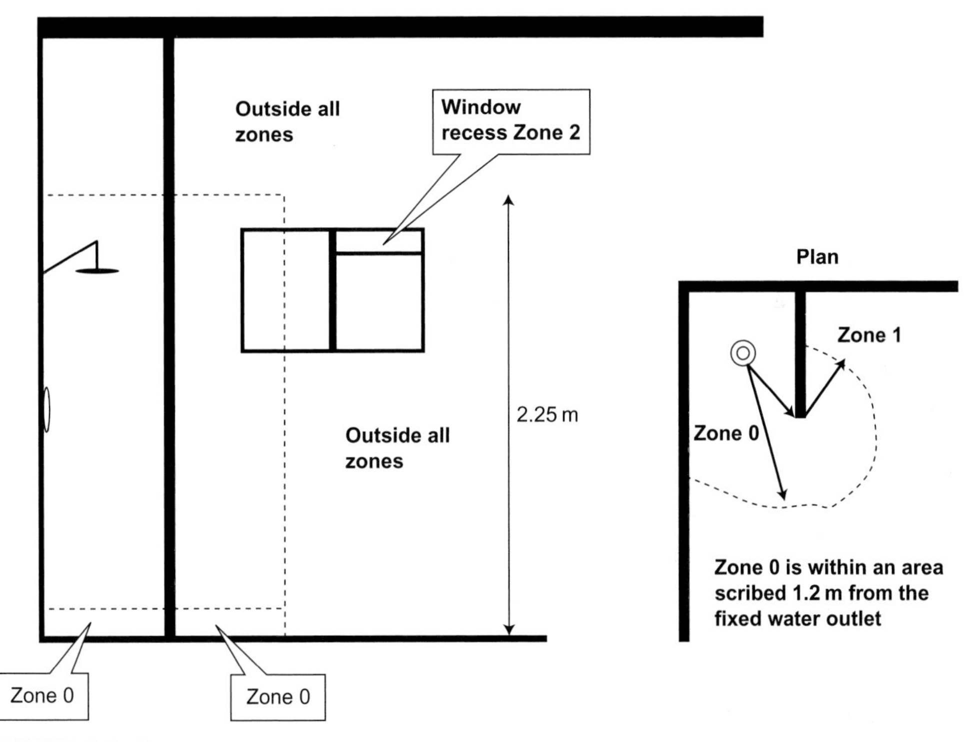

FIGURE 7.3 **Bathroom zones**

Where ADS is used all circuits within the bathroom or shower room must have additional protection provided by a 30-mA RCD as stated in regulation 701.411.3.3.

In the past, it has been a requirement that supplementary equipotential bonding was provided within these locations. In the 17th Edition of the Wiring Regulations, we have been provided with some options.

Supplementary bonding may be omitted provided the following conditions are met.

1. The installation within the building must have a protective equipotential bonding system.
2. All circuits within the bath/shower room are protected by ADS.
3. All circuits within the bath/shower room have 30-mA RCD protection.
4. All metalwork which could become live in the event of a fault or provide a path to earth within the bath/shower room is effectively connected to the protective equipotential system. This type of metalwork is known as an extraneous conductive part; it could include taps, metal baths, radiators, towel rails and various other items. If any of these parts are connected using plastic plumbing materials, then it is more than likely that they will not need bonding. In many instances, the extraneous conductive parts will be connected to the equipotential bonding via the plumbing system and CPCs of other circuits such as immersion heater and boiler circuits.

Remember, all four of these requirements need to be met before supplementary bonding can be omitted. Confirmation of the first three items should be simply a matter of a visual check and within a new installation these conditions should certainly be met.

Confirmation of the fourth requirement will need some simple testing to be carried out which will enable us to decide whether or not supplementary bonding is required.

A test between a known earth (exposed conductive part) and any equipment that you are not sure about must be carried out using a low-reading ohm-meter with long leads.

If the circuits within the location are protected by an RCD, then the formula to find the maximum resistance permissible is:

$$\frac{50}{I_{\Delta n}} = R$$

50 V is the maximum touch voltage, $I_{\Delta n}$ is the trip rating of the RCD and R is the maximum resistance permissible between exposed and extraneous conductive parts.

Example

$$\frac{50}{0.03} = 1667\,\Omega$$

Where there is no RCD protection the formula uses the current which will cause automatic disconnection of the protective device, I_a in place of $I_{\Delta n}$.

External influences, switchgear and current-using equipment

As with any installation we have to consider external influences; any current-using equipment and switchgear installed has to be suitable for use in the area where it is

Example

Where the protective device is a BS EN 60898 type B 6-A circuit breaker:

$$\frac{50}{30} = 1.66\Omega$$

The easiest way to find the current which would cause automatic disconnection of any protective device is to look at the tables in appendix 3. For example, if the circuit was protected by a 30-A BS 3036 semi-enclosed rewirable fuse, we need to look at table 3.2A and see that the minimum current required to operate the fuse is 87A.

It must be remembered that the current used is that which is required to operate the protective device within 5s. Where circuit breakers are used the time is not a consideration as they will always operate within 0.1s provided that they are installed to comply with the requirements of chapter 41.

Within most bath/shower rooms there will be more than one circuit. As electricians we must always look for the worst case scenario; for this reason the value I_a that is used in this calculation must be the highest within the area which may require supplementary bonding. We need to be careful as where circuit breakers are used I_a may be greater for a 20-A circuit breaker than it is for a 32-A circuit breaker. For instance, a 20-A type B will require a maximum of 100 A to ensure correct operation, whereas a 32-A type B will require 160A maximum for correct operation. However, if the circuit breaker is a 20-A type C it will require 200A to operate it. Now if we carry out the calculations we can see which is the lowest acceptable resistance that, will be required between exposed and extraneous parts to ensure the correct level of safety.

20A type B:

$$\frac{50}{100} = 0.5\Omega$$

32A type B:

$$\frac{50}{160} = 0.31\Omega$$

20A type C:

$$\frac{50}{200} = 0.25\Omega$$

Clearly, the 20-A type C requires the lowest resistance to ensure correct operation. If we use this value to confirm the quality of the bonding we will be conforming with the requirements of the regulations.

Having decided on the maximum value of resistance which is permissible between exposed and extraneous conductive parts, we have to carry out tests to find out whether supplementary bonding is required or not.

If there is no RCD protection, then the test is quite straightforward. It is simply a matter of testing between a known earth and any metal part that we think may need bonding. The instrument to be used is a low-resistance ohm-meter with long leads (method 2). It is very important to null the leads or record the resistance of the leads and subtract the value from the measured result. Once we have a measurement all that is required to do is compare the result with the maximum resistance. Provided our measured resistance is lower than the maximum permissible then no supplementary bonding is required.

All installations which have been installed since 2008 must have RCD protection for bathroom circuits; this of course is not the case where the installation was installed before 2008. Where the installation has RCD protection, the measurement between extraneous and exposed conductive parts is a little more involved as higher resistance values may need to be measured. The test procedure is exactly the same. The problem we have is that most of the low-resistance ohm-meters which we use will measure up to 100 Ω. Where RCD protection is in place, it is possible that we need to measure up to 1667 Ω, and this is where the difficulty lies because our low-reading ohmmeter will not measure high enough. The other option is to use an insulation resistance test instrument, but of course the value of 1667 Ω would be too low for this type of instrument as it would need to show a value of 0.0016 MΩ.

A practical solution would be to treat the situation as if we were trying to ascertain whether or not the metalwork was an extraneous conductive part or just a piece of metal. This would require the use of an insulation resistance test instrument, and we would need to measure between a known earth and the metalwork which we are unsure of. If the measured value was greater than 0.02 MΩ, then we would treat it as a piece of metal which could not produce a potential difference in the area and it would not require bonding. If however the measured value was below 0.02 MΩ, then in the absence of a more accurate measurement bonding should be carried out as this would ensure compliance with BS 7671.

If we decide that supplementary bonding is required, then we can carry out this bonding inside the room, or if this is difficult, maybe for visual reasons, it is quite acceptable to carry out the bonding in close proximity to the room, as close as possible to the point of entry of the pipework to the bathroom. An airing cupboard is a very good place to carry out any bonding, provided it is reasonably close to the bathroom. We need to be aware of any plastic pipework or connectors being used as they could affect the integrity of the bonding. There is no need to bond the copper legs supplying taps or radiators, where they have been used on a plumbing installation which is of plastic and the legs are installed through the floor for aesthetic purposes. There would also be no reason to bond a metal bath which is plumbed with plastic pipes.

installed, because bathrooms have specific zones – the regulations provide us with the minimum requirements suitable for these zones.

Zone 0

Any equipment installed in Zone 0 will be intermittently submersed in water and the classification for this area is IPX7. In general terms no electrical equipment should be installed in Zone 0; the exception to this would be a luminaire through the side of a bath, in which case it must be permanently connected and protected by SELV at 12 V a.c. or 30 V d.c.

Where SELV is used for protection in any zone, it is a requirement that the source for SELV is outside all zones.

Where the shower is a wet room with the shower outlet in the floor, Zone 0 is from floor level to a height of 100 mm.

Zone 1

This is the area immediately above the bath or shower tray to a height of 2.25 m from the floor level. The only switches permitted in this zone are those which are being

TABLE 7.1 IP requirements for zones

Zone	Minimum IP rating
0	IPX7
1 and 2	IPX4
Any zone other than Zone 0 where water jets may be used	IPX5

used to operate SELV circuits up to 12 V a.c. or 30 V d.c. Current-using equipment can be installed in Zone 1 provided that it is fixed or permanently connected and that it is suitable for the zone. The manufacturer's instructions for this equipment should be read thoroughly to ensure that it is suitable. The equipment should also be installed to comply with the manufacturer's instructions as well as any regulations. The minimum degree of protection for this area is IPX4, which is protection from splashing.

Zone 2

This area extends horizontally to 600 mm from Zone 1 and to a height of 2.25 m from the floor level. With the exception of shaver sockets to BS EN 61558-2-5 the only switches, socket outlets or accessories incorporating switches which are permitted in this area are those protected by SELV at 12 V a.c. or 30 V d.c. The minimum degree of protection permissible for this area is IPX4.

All other areas within the bathroom or shower room are classed simply as outside all zones. In general terms, this area is above a height of 2.25 m from the floor in Zones 0 and 1 and any area 600 mm horizontally from Zone 1 at any height.

Socket outlets at low voltage are permitted in the bathroom provided they are at least 3 m horizontally from Zone 1.

Some product manufacturers provide drawings with the equipment which they supply; these drawings often show zones, particularly around basins. Where a manufacturer states that an area is a zone, even if it is not rated as a zone in part 701, we have to comply with the manufacturer's recommendations. This will ensure compliance with regulation 134.1.1 (Table 7.1).

Electrical underfloor heating systems

Any electrical underfloor heating system must be suitable for the environment in which it is installed and comply with the required product standard. If we install a heating element that is designed for underfloor heating in a bathroom and it complies with the required British Standard, it will be compliant with BS 7671. In addition to this, we must ensure that any heating element installed under the floor in bathrooms is encased in a metal sheath or enclosure, or it can be covered with a fine metallic mesh. The sheath, enclosure or mesh must be connected to the CPC of the circuit supplying the element. If the heating is supplied by SELV, then the metallic covering is not required.

Before installing any underfloor heating, it is important that the positioning of any equipment such as sanitary ware, baths, shower trays or cupboards is known. This is because any heat from elements installed under the floor must be allowed to dissipate through the floor. If any equipment were to be installed over the element, it could overheat and would present a potential fire hazard.

It must also be remembered that part 701 is for bath and shower rooms, where underfloor heating is to be installed; the requirements of part 753 should also be referred to.

SECTION 702

Swimming pools and other basins

This section applies to swimming pools, paddling pools and fountains along with the areas around them. To identify the correct level of protection, the areas are zoned.

Zone 0 is inside the swimming pool, paddling pool or fountain. Protection must be by SELV limited to a voltage of 12 V a.c. or 30 V d.c.

Any equipment installed inside Zone 0 must have a level of protection IPX8 and be suitable for installation within the zone. A typical item of equipment would be an underwater luminaire. The only wiring system permitted within this zone is one which is supplying equipment within the zone, and any metallic covering of the wiring system must be bonded.

Where equipment is installed in Zone 0 of a fountain, it must be mechanically protected to AG2 – Georgian wired glass or mesh would be suitable for this. The wiring to equipment in Zone 0 of a fountain must be installed using the shortest possible route, and must be mechanically protected to AG2, which is medium severity, and have an IP rating of IPX8. Where the cable is to be installed at a depth of 10 m or more, the cable manufacturer must be consulted as to the suitability of the cable.

Zone 1 is the area around the pool or fountain; it extends from Zone 0 for a distance of 2 m vertically and 2.5 m horizontally.

Where there is equipment such as a slide or diving board, Zone 1 extends vertically around the equipment for a distance of 1.5 m and horizontally to a height of 2.5 m. Protection against electric shock for this area is the same as Zone 0 and the minimum IP rating for equipment in this zone is IPX4, or IPX5 if water jets are used for cleaning purposes.

Where switchgear and control gear have to be installed for a swimming pool, common sense has to be used. In general, it is not permitted to install switchgear or control gear in Zone 0 or 1; however, in some installations there may not be a Zone 2 and in these instances socket outlets and switches are allowed in Zone 1 provided they meet the following requirements:

1. There is nowhere else to locate them.
2. They are installed at least 1.25 m from the border of Zone 0 and at least 300 mm from the floor level.

There are three options for protecting sockets and switches where they are installed in Zone 1.

1. They are supplied by SELV not exceeding 24 V a.c. or 60 V d.c. with the SELV source being installed outside Zones 0 and 1.
2. They are protected by a 30-mA RCD (maximum) with a disconnection time of 40 ms at a fault current of 150 mA. The circuits must also have overload and fault protection.
3. Electrical separation can also be used to supply one item of equipment, with the source of the separated circuit being located outside Zones 0 and 1.

These rules also apply to socket outlets and switches installed in Zone 2. The only difference is that the source for SELV or separation must be installed outside Zone 2.

Fixed current using equipment in Zone 0 or 1 is not permitted unless it is designed and constructed specifically for use within these zones. Other equipment which is intended only to be used when people are outside Zone 0 is permitted, provided it has the same protection as circuits supplying socket outlets and switches as described earlier.

Underfloor heating is permitted provided that it is embedded in the floor if it complies with **one** of the following three requirements.

1. It is protected by SELV, provided that the source for the SELV is installed in Zone 2 if protected by an RCD or outside all zones if not RCD protected.
2. The heating incorporates an earthed metallic sheath which is bonded to all exposed conductive parts within the location, and protected by an RCD with a maximum $I_{\Delta n}$ of 30 mA with a disconnection time of no greater than 40 ms.
3. It is covered by an embedded earthed metallic grid which also meets the requirement of point 2.

A good rule of thumb is to ensure that any equipment installed in or around a pool or fountain is purposely made for the job and that the manufacturer's instructions are followed to the letter, provided the equipment is compliant to the required British or European Standard.

SECTION 703

Rooms and cabins containing sauna heaters

This section of the regulations applies to any room containing a sauna heater; the reasons for a sauna being a special location are the same as for bathrooms and swimming pools. In most cases, people using these areas are partially clothed and wet; this increases the risk of electric shock considerably, due to the possibility of greater areas of skin contact with any extraneous conductive part, and the reduced resistance of the skin due to its being wet.

Only electrical equipment which is strictly necessary for the safe operation of the sauna is permitted in the room containing the sauna. This equipment would be the sauna heater, its thermostat control and a luminaire, which must be suitable for use

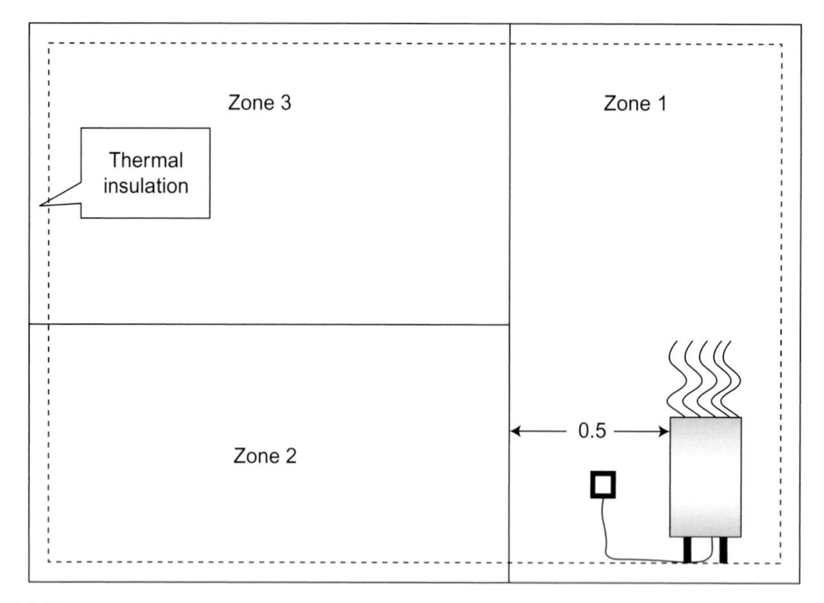

FIGURE 7.4 **Sauna zones**

within the environment. All other equipment such as light switches must be installed outside the sauna.

All equipment must be protected to a minimum of IPX4; if water jets are to used to clean the area, then the IP rating must be increased to IPX5.

Zone 1 extends from floor to ceiling, and 500 mm around the sauna heater or the cold side of the wall insulation, whichever is the closer. Only the sauna heater is permitted in this area.

Zone 2 extends to 1 m above the floor and includes the whole of the area outside Zone 1 to the cold side of the wall insulation. Only equipment which is part of the sauna heater may be installed in Zone 2. An exception would be a piece of equipment which has been manufactured specifically for installation in saunas and is installed in accordance with the manufacturer's instructions.

Zone 3 extends from 500 mm above the floor level and includes the whole of the area outside Zones 1 and 2 to the cold side of the wall/ceiling insulation. Any equipment in this zone must be able to withstand 125°C and any wiring must be able to withstand a minimum of 170°C (Figures 7.4 and 7.5).

In reality, it is always better to install any wiring on the cold side of the thermal insulation; if this is not possible then the wiring in Zones 1 and 3 must be heat resisting. This is not a requirement for Zone 2 as it is not expected that high temperatures will be present in this zone. Where it is only possible to install the wiring system within the sauna, any metallic sheath or conduit must not be accessible during normal use.

Protection for circuits in a sauna can be by SELV or ADS. Additional protection must also be provided for circuits by the use of an RCD. However, RCD protection need not be provided for the heater elements unless it is recommended by the manufacturer.

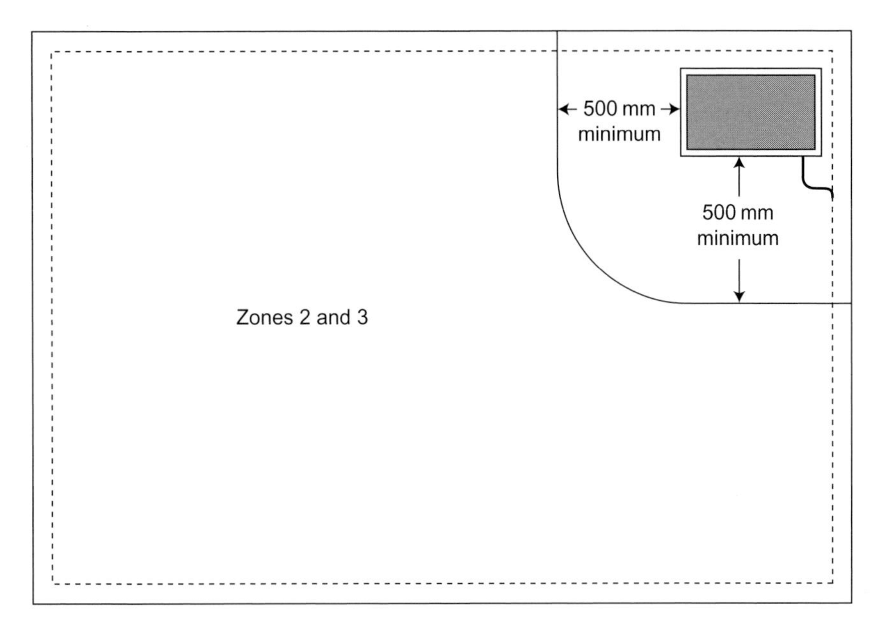

FIGURE 7.5 **Sauna zones**

SECTION 704

Construction and demolition sites

This section of the regulations applies to the actual construction and demolition areas and not to offices, toilets or other buildings used by people when not actually working in the construction and demolition areas.

These sites are usually very harsh environments and the risk of electric shock is high. This is due to the areas often being open to the weather, and lots of extraneous conductive parts which are virtually impossible to bond. Other areas of concern are damage to trailing cables and the necessary use of hand tools.

TN-C-S systems must not be used for construction and demolition site supplies. Where a TN-C-S supply is provided by the supply provider, it must be converted to a TT system; this is easily achieved by isolating the supply earth from the installation and installing an earth electrode for use with an RCD (Figure 7.6).

All socket outlets and any circuit supplying hand-held equipment up to and including 32 A must be protected by one of the following methods.

- SELV or PELV.
- ADS with additional protection provided by a 30-mA RCD with a maximum disconnection time of 40 ms at 5 times its rating.
- Reduced low voltage. This is usually 110 V between lines and is supplied by a centre tapped transformer. The disconnection time for these circuits is a maximum of 5 s, with the maximum Z_s values for the circuits being found in table 41.6.
- Electrical separation, with each socket outlet being supplied by a separate transformer.

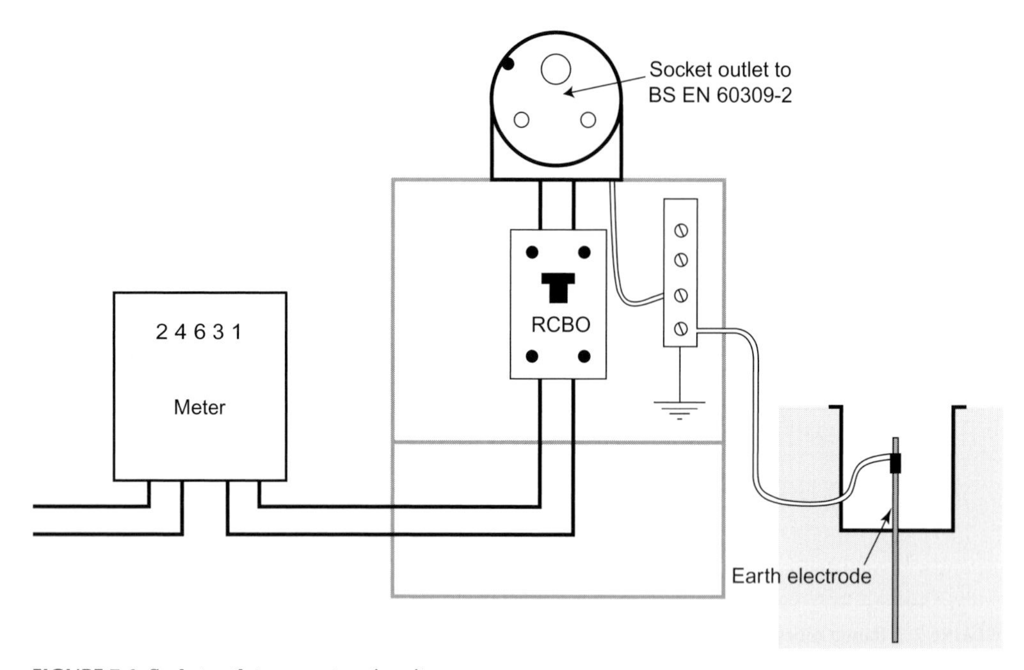

FIGURE 7.6 Socket cutlet on construction site

Any socket outlets rated at 16 A or more must be to BS EN 60309-2 (industrial socket outlet). All assemblies used on construction and demolition sites must be compliant with BS EN 60439-2, which gives the particular requirements for assemblies for construction sites.

Where socket outlets are used with a rated current of greater than 32 A, the circuits must be protected by an RCD with a maximum rating of 500 mA. Disconnection times must be as shown in table 41.1. This table requires that the RCD must operate within 0.2 s.

Due to the type of work which is often carried out on construction sites, it may be necessary to consult section 706, which is for conducting locations with restricted movement.

The inspection of the electrical installation should be carried out regularly, with the maximum period between inspections being 3 months. RCDs must be tested daily or before use if not in use each day; this can be carried out by an ordinary person by using the test button. Formal testing of RCDs should be carried out by a responsible person every 3 months using an RCD test instrument.

SECTION 705

Agricultural and horticultural premises

As with all other special locations, these areas have their own particular risks. Electrical installations and equipment in these areas are generally exposed to a very harsh environment, and can be prone to mechanical damage and corrosion. Rats and mice are a major problem on many farms and can lead to fire risks, particularly where straw and hay are stored; and of course many farms have livestock, which also needs to be protected from electric shock.

TN-C-S systems must not be used for agricultural and horticultural supplies. Where a TN-C-S supply is provided by the supply provider it must be converted to a TT system. This is easily achieved by isolating the supply earth from the installation and installing an earth electrode for use with an RCD.

ADS must be used throughout the installation, although the disconnection time of 0.2 s has been replaced by the use of RCDs.

Socket outlet circuits not exceeding 32 A must be protected by an RCD with a maximum operating current of 30 mA.

Socket outlets with a current rating of greater than 32 A must be protected by an RCD with a maximum current rating of 100 mA. All other circuits must be protected by an RCD with a maximum operating current of 300 mA. This rating of device will also provide the required protection; obviously RCDs with a lower rating are also permitted for fire protection. Where a 300-mA RCD is used to protect a distribution circuit which supplies other circuits, a time delay or S type may be used. This will provide improved continuity of the supply in the event of a fault on a single circuit.

Where an earth electrode is to be used consideration must be given to the electrode resistance. In most cases, we accept that the electrode resistance must be no greater than 200 Ω; however, where an electrode is used for any RCD with a rating greater than 200 mA the formula $R_A \leq 50/I_{\Delta n}$ is applied.

Example

Where a 300-mA RCD is to be used for fire protection:

$$\frac{50}{0.3} = 167\,\Omega$$

This would be the maximum permitted resistance of the electrode and the protective conductor connecting it to any exposed conductive parts.

Electrical equipment

Very careful consideration must be given to the positioning of all electrical equipment in these types of environment. Fire is a real danger where heaters are installed. Radiant heaters must be positioned with a clearance of at least 500 mm from any combustible surface unless the manufacturer's information indicates otherwise. All equipment must be IP rated to a minimum of IP44, and where this is not possible the equipment must be installed in an enclosure which complies with IP44. Damage by impact must also be considered: it is common sense to keep electrical equipment well out of reach of animals and livestock as well as vehicles and machinery. Although often it is impractical to keep the wiring system completely out of harm's way, in these situations the wiring system must be selected and installed to minimise any damage that may be caused. The degree of impact protection required by BS 7671 for this type of installation is 5 J. For information purposes only, this is a mass of 1.25 kg dropped from a height of 400 mm.

It is a requirement of the regulations that documentation is provided by the installer. This documentation must include the following information:

- A simple drawing showing the distribution system
- The routing of all concealed cables
- A plan showing the location of all of the electrical equipment
- A diagram showing the locations of all of the bonding connections.

Where heating appliances are used they must have a visual indicator showing clearly the on and off position; a neon indicated switch is usually the best method as it can be seen without having to inspect it closely.

Isolation

Most horticultural and agricultural enterprises consist of more than one building. Each building must have its own means of isolation which must be clearly marked. Where circuits are only used occasionally the isolation must include the neutral as well as all other live conductors.

Protective bonding

Due to the nature of the environment, it is very important that care is taken over the selection of the type of bonding conductors used. Any suitable material can be used, but it must be able to withstand any mechanical damage and corrosion that is likely to occur.

SECTION 706

Conducting locations with restrictive movement

This type of location includes any conducting location where movement is restricted. An example of this would be a steel vessel used in a food processing plant or any other conducting equipment which due to the nature of its use would require a person to work inside it occasionally. In the event of an electric shock the person inside the location would have little or no chance of being able to move away from the shock.

By far the best method of protection from electric shock in this type of environment is the use of battery-powered equipment, although sometimes this may be impractical due to the nature of the work being carried out.

For hand-held equipment, electrical separation or SELV must be used. Fixed equipment can be supplied by SELV, or as long as all extraneous conductive parts and exposed conductive parts are bonded to the system earth PELV may be used. Electrical separation may also be used.

If it is required that the supply voltage is used, then Class II equipment can be used provided the circuit supplying it is protected by a 30-mA RCD.

Where it is not possible to use Class II equipment, and shock protection is simply by ADS, supplementary bonding must be installed, and must connect all exposed conductive parts of the equipment to all conductive parts of the location.

As you can see, where the supply voltage is to be used, bonding forms a vital part of the protection.

SECTION 708

Electrical installations in caravan and camping parks

This section of the regulations applies only to the supply to a caravan, tent or recreation vehicle, not to the actual vehicle or any permanent buildings on the site.

The greatest problem with this kind of installation is that it is impossible to provide equipotential bonding; this is because the persons using the site spend a lot of time in contact with the actual earth itself.

TN-C-S supply systems are not permitted to be used on caravan or camping sites. This is because of the danger which could be present if the supply PEN conductor became open circuit. Consider the situation where a caravan is connected to a socket outlet which is supplied from a TN-C-S system. If the PEN conductor was to become disconnected, the outer skin of the caravan may become live, and a potential difference would be present between the caravan and earth. This would present a very dangerous situation indeed (Figure 7.7).

Supplies to the camping pitches must be from either TN-S or TT systems. In the event of a TN-C-S supply being present at the main intake, it must be converted to a TT system. This is achieved by isolating the earth at each pitch and installing an RCD (Figure 7.8).

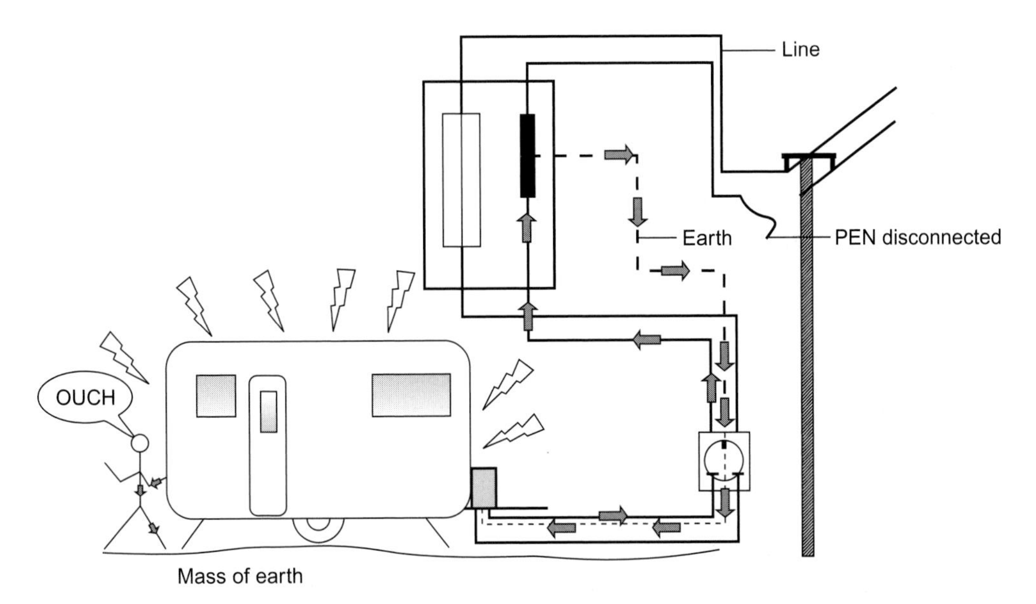

FIGURE 7.7 **TN-C-S system on a caravan site**

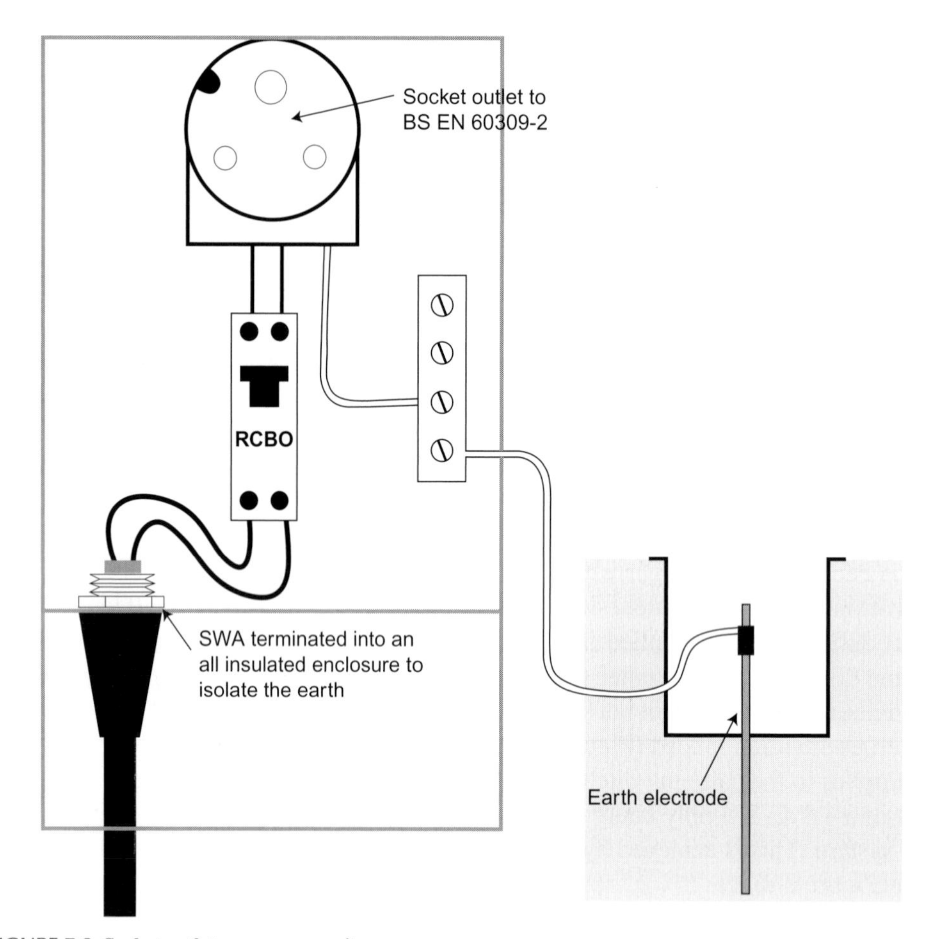

FIGURE 7.8 **Socket outlet on a caravan site**

Socket outlets must be suitable for the environment, as must any other equipment used in any electrical installation. They must also be of the industrial type (BS EN 60309-2) rated at a minimum of 16 A and be protected to a minimum of IP44. Each socket outlet must have its own overcurrent and RCD protection.

Each pitch must have a minimum of one socket outlet with a maximum of four socket outlets being grouped in one location. The outlets must be a minimum of 500 mm and a maximum of 1.5 m from ground level. Where flooding or high levels of snowfall are likely the height may be increased as required. In all circumstances, the socket outlet must be no greater than 20 m from the pitch.

The supply to the outlets can be either underground or overhead, although underground is the preferred method. Where the cables are buried they must be a minimum of 600 mm deep and placed away from any areas where tent pegs may be used.

Overhead cables must be a minimum of 3.5 m from the ground with the height being increased to 6 m where vehicle movement is likely.

SECTION 709

Marinas and similar locations

This section is intended for use on supplies to pleasure craft when moored in marinas; it is not intended for the actual electrical installation of the pleasure craft. The installation on board pleasure craft is covered by its own set of British Standards (BS EN 60092-507).

A pleasure craft is a boat, yacht, motor launch or any floating class or vessel which is used for sport or leisure.

House boats which are permanently moored must comply with BS 7671 in its entirety.

As with many other special locations, the use of TN-C-S supply systems is not permitted for the supply to pleasure craft and they must be converted to TT systems. This does not apply to any permanent buildings in the marina. TN-S systems may be used to supply pleasure craft and any permanent buildings.

The problems encountered for marinas are corrosion, the movement of the pontoon and jetties as well as the craft, along with the lack of an equipotential zone. Due to the environment, consideration must be given to the selection of non-corrosive materials and protection must be to a minimum of IP3X for dust and solid foreign bodies; and due to the presence of water, the best possible protection should be provided. I would always provide protection to IPX6 due to the changing conditions in a marina. Lesser protection may be provided where you can be certain of the type of influence.

Protection from impact should also be provided where there is likely to be any contact between the electrical installation and any other equipment such as trolleys used to transport equipment to and from boats.

The supply voltage to pleasure craft must not exceed 230 V single phase and 400 V three phase. The supply must be RCD protected at the marina and the pleasure craft. Isolating transformers may also be used as long as each secondary winding supplies only one pleasure craft.

Socket outlets used for the supply of pleasure craft must be industrial type outlets to BS EN 60309-2. Unless it is known that pleasure craft using the marina will have a high current demand, 16 A outlets should be used with an IP rating of IP44 and a key position of 6 hours.

It is preferable that socket outlets installed on the same jetty are on the same phase; where this is not possible they must be positioned so that they cannot be touched at the same time by the same person. It is not unusual for socket outlets to be installed in pillars similar to those used on caravan sites. In these cases, a notice warning users of the maximum voltage present must be positioned so that it can be clearly seen.

A notice must be provided which includes the text as shown in BS 7671. This can be positioned next to the socket outlets used for the supply to the pleasure craft or handed to each berth holder on arrival.

In other areas of the marina, the installation must be suitable for the environmental conditions, particularly with regard to the selection of the type of materials being used and the location of equipment. All parts of the installation must be inspected and tested regularly, and records of the inspection and test must be kept.

SECTION 711

Exhibitions, shows and stands

This section is for use with any temporary installations used for exhibitions, shows, displays or stands. These can be indoors or outdoors; however, it must be remembered that the section only applies to the actual temporary installation itself, not the installation of the building from which the supply may be taken. Also it does not apply to mobile units.

The problems surrounding this type of installation are that it is temporary, and because of this it is not usually fixed to a permanent structure, this in turn could result in mechanical stress due to movement of the structure. Added to this is the presence of the general public.

Due to the nature of this type of installation, the wiring system must be well protected from mechanical damage. Where there is access to the general public, flexible cords must be very well protected against mechanical damage. Where the building which is being used for the show or exhibition stand does not have a fire-alarm system, flame-retardant cables, conduit or trunking must be used, and cables should not be joined unless it is in an accessory. The trunking and conduit must have protection to a minimum of IP4X.

Switchgear and control gear must be placed in an enclosure which can only be opened by the use of a key or tool; this is unless they are intended to be used by an ordinary person. The means of isolation for the show or exhibition stand must be readily accessible to the user of the stand. This section of the regulations also states that 'where an electric motor may give rise to a hazard it must be provided with an effective means of isolation'. I would never consider putting an electric motor anywhere without providing a means of isolation which is as close to it as is sensible and accessible.

TN-C-S systems must not be used and must be converted to a TT system; TN-S systems may also be used.

The supply to the temporary structure must be protected by a time delay or S-type RCD which has a maximum rating of 300 mA. This is to provide discrimination for final circuits. All other circuits apart from emergency lighting circuits must be protected by a 30-mA RCD.

All metal parts which can be accessed from within the stand must be bonded to the main earthing terminal within the unit.

As with all other electrical installations, consideration must be given to heat. This could be generated by the electrical equipment and could cause danger, damage or fire to materials which surround it; or it could be an external source of heat which could cause similar damage to the actual electrical installation.

Socket outlets must be installed to reduce the use of trailing leads. Where socket outlets are installed in the floor they must be protected from the ingress of water and mechanical damage from persons or equipment passing across them.

These installations must be inspected and tested before being put into use and then on a regular basis. Records should be kept of these tests.

SECTION 712

Solar PV power supply systems

The use of PV systems for generating energy in dwellings is a relatively new idea in the UK; however, with the government promoting the use of renewable energy you can be sure that in a very short time PV systems will become very popular indeed.

Before we start looking at the requirements of BS 7671, let us take a look at what a PV system is, and the types of configurations which we can choose from when a solar electricity installation is to be installed.

Once the system has been installed the sunlight is free and the maintenance costs are very low. The energy can also be stored.

It is very unlikely that a PV system will generate enough electricity to completely power the average modern home, although if the energy is stored it can be used as and when required. There are four different ways in which PV installations can be utilised.

Stand-alone system

This is the simplest type of PV system and can be very cost-effective, particularly for providing power to a shed or small remote building. In this type of system, the electricity produced by solar energy is stored in a battery and used as required. To prevent the battery from becoming overcharged a controller is required; this will manage the flow of electricity into the battery and will also prevent the battery from becoming completely discharged. This is a very important piece of equipment as overcharging or completely discharging a battery will dramatically reduce its life.

The energy stored can then be used to supply a d.c. load directly from the battery. The most common voltage provided by a solar panel is 12 V d.c. The power output is dependent on the panel size. A number of PV panels can be connected in parallel or series to provide the desired voltage as shown in the following example.

Four 12-V 30-W panels connected in parallel will provide a 12-V supply with an output of 120 W. The same panels connected in series will provide a 48-V supply with an output of 120 W. The same arrangement can be used for batteries, the difference being that batteries in parallel will provide a higher current. For this reason, a mixture of batteries connected in series and parallel may be used (Figure 7.9).

Where a 230-V a.c. supply is required, an inverter can be connected to the storage batteries; this will then convert d.c. to a.c. and increase the voltage as required.

This type of system is relatively cheap and easy to install and provides a supply where other methods may be difficult.

Grid tie and grid tie with battery backup

A grid tie system allows the energy provided by the PV panels to be fed back into the national grid via a grid tie inverter; the supply company will then pay you for the energy which has been provided by your system. In a grid tie system, the greater the

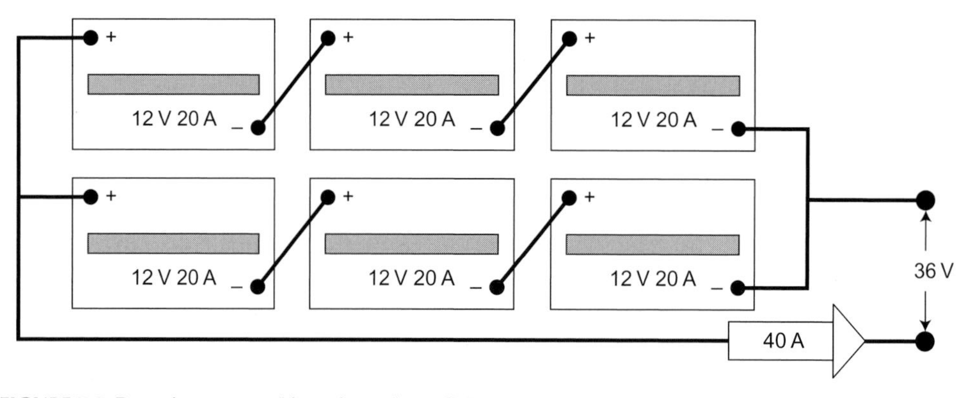

FIGURE 7.9 **Batteries connected in series and parallel**

voltage provided the more efficient the system will be. Multiple PV panels are often connected in series to provide a greater voltage. This type of system will require the cooperation of the electricity supply company as a special import/export meter will need to be installed.

Depending on there being an area suitable for battery storage, this system can be enhanced by adding a battery backup, hence the name grid tie with battery backup.

This system first charges the battery bank and keeps it topped up. Any surplus energy is fed back into the grid via a grid tie inverter. In the event of a power cut, the battery storage can be used via an inverter to provide a supply to any chosen circuits.

Where the system includes a provision for energy storage, a controller must be installed to allow the energy to charge the batteries; this controller will prevent the batteries from overcharging and becoming damaged. It will also prevent the batteries from being discharged to a level where the battery life is reduced.

This type of system can also be expensive to install.

Grid fallback system

This type of system is relatively inexpensive to install; however, it does require a storage area for batteries. With this type of system, the energy collected from the PV panels charges a bank of batteries. The energy stored in the batteries is then used to supply selected circuits via an inverter. In the event of the battery bank running flat, the system will automatically switch the circuits back to the main supply until the batteries are sufficiently charged to be used again.

This system has the benefit of your being able to use your own power when you need it, thereby making a saving on your electricity bills. It provides most of the benefits of the grid tie system, but instead of selling your generated electricity back to the supply company you are using it yourself. A charge controller will also be required for this system.

Solar panels must be fixed securely, particularly where they are fixed onto the fabric of a building; great care must be taken to ensure that the panels are mounted slightly off the roof to allow free passage of water under them. They must also be able to withstand any wind pressure.

Another consideration is the additional load which will be placed onto the roof structure. Where there are a lot of panels it is usually a good idea to seek the advice of a structural engineer.

Planning is also required in some instances and the advice of the local authority should be sought if there is any doubt. Planning would not be granted for instance if the panel were to protrude 200 mm beyond the plane of the roof slope or wall, nor would it be permitted if it were to protrude above the highest part of the roof. The chimney is not considered part of the roof.

When installing solar panels it is important to ensure that they are sited correctly; this will ensure that the maximum efficiency is gained from the installation. PV panels need light to operate; the greater the light, the better performance will be achieved.

Not only does the sun move across the sky, it is also at a different angle during different months of the year. For instance, the sun's angle in London in mid-summer will be 62.5° and in mid-winter it will be 15.5°; this will make a difference to the energy which the PV array will collect. Solar tracking devices are made which will track the sun throughout the year, but this adds to the expense of the installation, and remember PV arrays still work even if not in direct sunlight, although direct sunlight is as good as it gets.

Shading is a problem for PV panels and should be avoided if at all possible. When a part of a panel is shaded not only does it not generate, it also becomes a load on the system and can use energy from the other panels. A hot spot in the shaded panel can also be caused which in time will damage the panel. Where shading cannot be avoided, blocking diodes must be installed to prevent energy from other panels being fed into the shaded panel. This is not a major problem provided the correct precautions are taken at the time of installation.

Regulations

PV installations up to 16 A per phase which are connected in parallel with the public low-voltage distribution network must comply with engineering recommendation G83/1 (2003). Larger installations must comply with engineering recommendation G59/1.

For an installation which is greater than 16 A, written approval must be obtained from the relevant distribution network operator before commencing work. All installations must comply with BS 7671.

The circuits which are being supplied by the PV installation must always comply with BS 7671 where the system includes a 230-V inverter. BS 7671 does not cover requirements for a stand-alone system which operates from an extra-low-voltage battery pack. However, even under these circumstances great care should be taken as fire is always a risk where electricity is present.

The first important thing to remember with regard to PV panels is that they are always live. Unlike solar water heating PV panels do not require heat. PV panels work on the amount of light which they are exposed to and the hotter they are the less efficient they become: heat can reduce the efficiency of a solar panel by as much as 10%.

TABLE 7.2 Abbreviations used in PV systems

STC	Standard test conditions
V_{oc}	Open circuit voltage
V_{mp}	Maximum power point voltage (this is voltage under load)
I_{sc}	Short circuit current
I_{mp}	Maximum power point current (this is current under load)

A 12-V PV panel will produce a much higher voltage, often in excess of 20 V when off load; for this reason great care must be taken when working with these panels. A good idea is to keep them well covered until ready for use, or failing that, ensure that the ends of the leads from the solar panel are joined together. This will prevent any nasty electric shocks, particularly when a number of panels are joined together, which of course is usual.

A switch disconnector must be provided to isolate the PV array from the rest of the system. To avoid electric shock, the PV array should be connected to the switch disconnector at the earliest opportunity. The switch disconnector must be double pole, and it is very important to ensure that the disconnector used to isolate the d.c. side of the system is rated for d.c. currents. When switching d.c. the current creates a much bigger arc and the switch has to be able to cope with this. It is permissible to use a double pole circuit breaker to BS EN 60898 as a disconnector. Where the installation consists of more than one string, isolation must be provided for each string, which must be readily accessible.

It is usually a good idea to provide isolation as close to the array as possible and also at the inverter end.

For PV systems which have no more than three arrays, overcurrent protection is not required; however, the cable from the arrays must be capable of carrying a minimum of 1.25 times the maximum short circuit current of the arrays. This must be calculated before the installation can begin to ensure that the correct cable is selected. All PV modules will have a data sheet or plate providing the information which is required to carry out the calculation. The data sheet shows values with abbreviations next to them (Table 7.2).

As an example let us take an array of six panels each with an I_{sc} of 7.7 A, and a V_{oc} of 22 V.

As they are in series, the I_{sc} would be 7.7 A; therefore, the cable would need to carry $7.7 \times 1.25 = 9.63$ A.

The V_{oc} would be $6 \times 22 = 132$ V (Figure 7.10).

If we were to connect the panels in two strings of three panels each, then the I_{sc} would be 15.4 A and the cable required would need a rating of $15.4 \times 1.25 = 19.25$ A.

The V_{oc} would be $3 \times 22 = 66$ per string. As there are two strings the total V_{oc} would be 132 V (Figure 7.11).

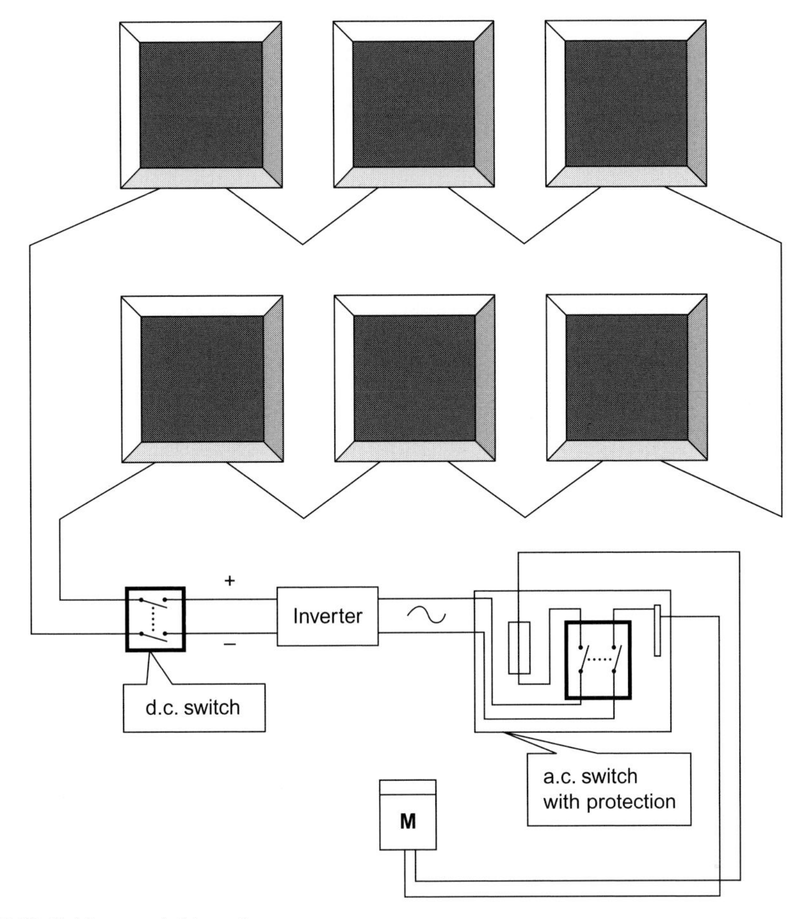

FIGURE 7.10 **Grid connected in series**

Where the installation consists of more than three arrays the cable must be fused; the fuses must be fitted to both the positive and negative of the string cables for all strings. The fuses must:

- Be rated for d.c. operation at the fault energies present
- Be rated for operation at $V_{oc\,(STC)}$
- Have a tripping current which is less than $2 \times I_{sc\,(STC)}$ or the string cable current capability, whichever is the lower value.

Where the system is being used to supply the grid, isolation must also be provided between the PV system (a.c. side of the inverter) and the a.c. mains.

Where an inverter is used for feedback into the grid, it is important to ensure that the correct type of inverter is used and that it is rated correctly; this is for safety and efficiency reasons.

The voltage and power rating of the inverter should match as closely as possible the voltage and power rating of the PV array as oversizing the inverter will lead to high power losses. To achieve the required voltages it is necessary to connect the PV

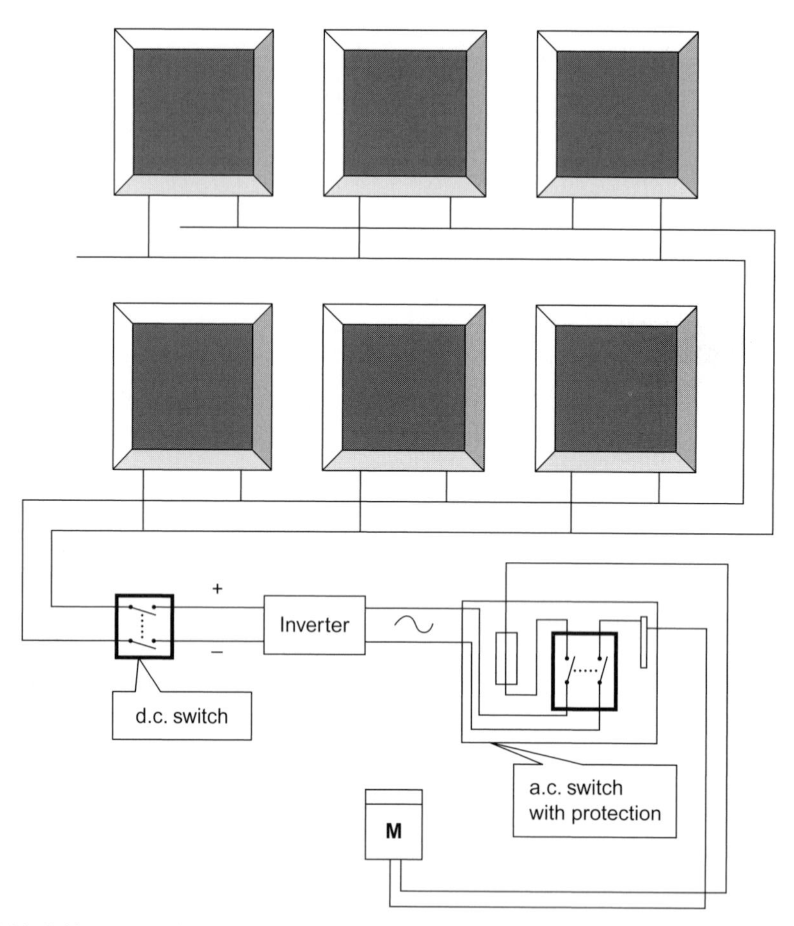

FIGURE 7.11 Grid connected in parallel with dedicated a.c. control

panels in series and parallel. As an example let us take a roof which due to its shape and size can accommodate 12 panels on two rows – the top row has seven panels and the bottom row has five. The panel characteristics at STC are:

P_{max}	110W
V_{oc}	22V
I_{sc}	7.7A
V_{mp}	16.9V
I_{mp}	7.1A

If the panels were connected in a string the V_{oc} would be:

$22 \times 12 = 264$ V, the total power would be 1320 W and the current would be 7.7 A (I_{sc}).

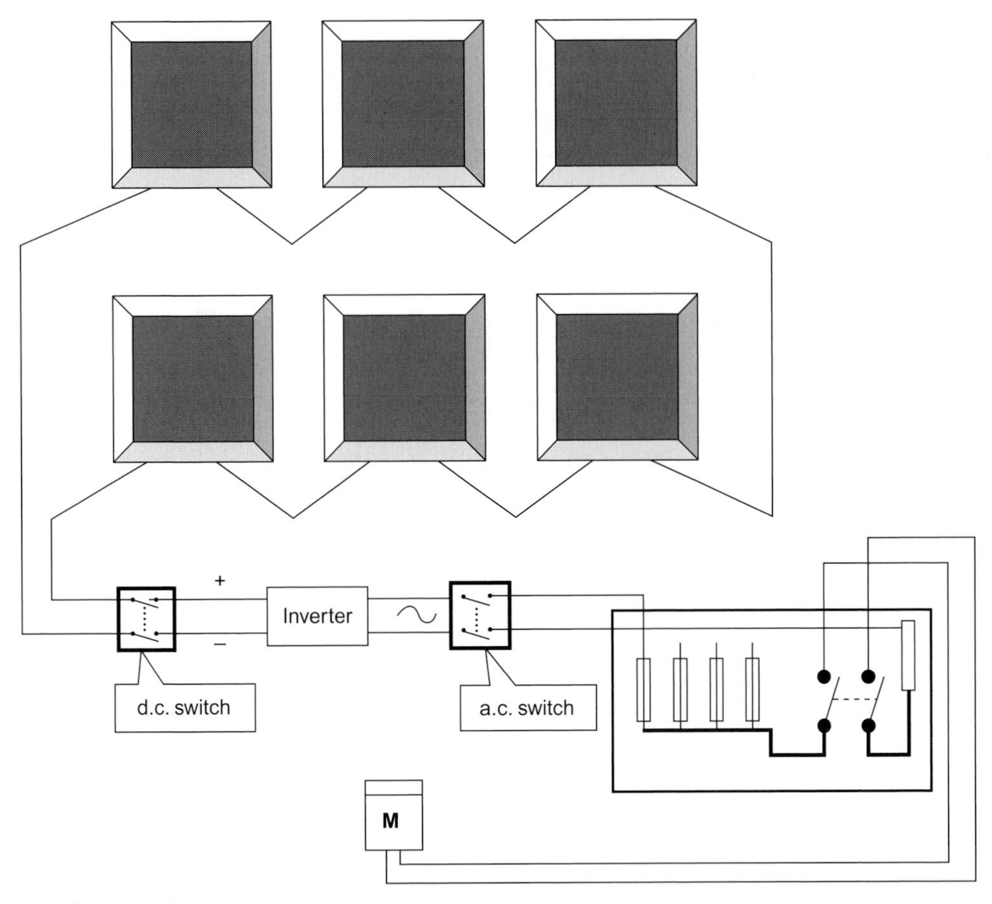

FIGURE 7.12 Grid connected through consumer unit

In some instances, it may not be possible to connect the panels in one string. If we were to connect the top row (string 1) we would have a V_{oc} of:

$22 \times 7 = 154$ V with a total power of $7 \times 110 = 770$ W and an I_{sc} of 7.7 A.

The bottom row (string 2) would also be connected as a string which would provide a voltage of:

$5 \times 22 = 110\ V_{oc}$, a power of $5 \times 110 = 550$ W and a total current of 7.7 A.

If we now connect the strings in parallel we will have a total voltage of:

$154 + 110 = 264$ V, a total power of $770 + 550 = 1320$ W and a total current of 15.4 A.

The inverter would need to be selected to suit whichever arrangement was preferred.

Where the preferred choice is for a grid tie system, there are a few important things to remember. First, it is possible to feed the installation from the solar panels via an inverter through a protective device in the consumer unit (Figure 7.12). This is fine although not ideal.

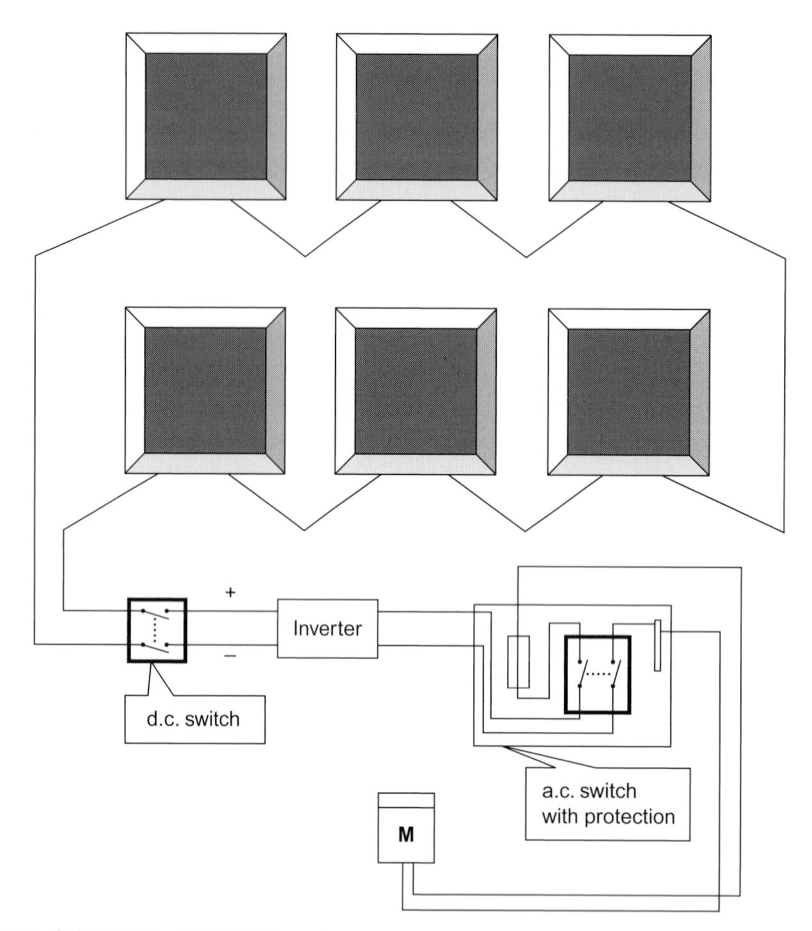

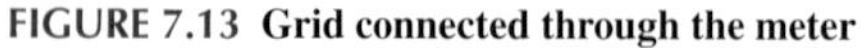
FIGURE 7.13 Grid connected through the meter

A far better option is to keep the PV installation separate and feed into the installation via the integrating meter. This would keep the PV installation separate from the existing installation, and would not require the installer to carry out a periodic inspection on the existing installation as it is not being touched. However, I would always recommend to the client that a periodic inspection or survey is carried out for safety reasons (Figure 7.13).

Batteries

Where batteries are being used as a backup (grid tie backup) or to regularly supply circuits (grid fallback), the batteries have to be calculated to suit the load; if not, the life of the batteries will be considerably reduced. Before we look at the sizing of the batteries, we need to decide which type of battery we will be using.

There are two types of battery which are popular for PV installations. They are both lead acid batteries: one is called a gel battery and the other is an absorbed glass mat (AGM). These batteries are well suited to PV installations as they hold their charge well, and do not degrade as easily as conventional lead acid batteries.

TABLE 7.3 Battery cycle life

Depth of discharge (%)	Number of cycles
10	1800
30	1400
50	100
80	350

Another advantage is that they give off very little hydrogen gas, so they can be stored in poorly ventilated places; they are also maintenance free. Lead acid car batteries should only be used as a last resort, as they are constructed to provide high currents for short periods, which is not what is required for PV systems.

Battery capacity is measured in ampere hours (AH). The AH rating shows for how many hours a battery will manage a specific drain. For example, a 200-AH battery will theoretically provide a current of 1 A for 200 hours, 4 A for 50 hours or 200 A for 1 hour. I say theoretically, as in reality the discharge power will reduce dramatically if the battery is discharged quickly. Another point worth mentioning is that all batteries have what is called a cycle life; most manufacturers will provide information showing the battery cycle life versus depth of discharge. The greater the depth of discharge, the shorter the battery life will be (Table 7.3).

From this table, it can be seen that it is far better not to discharge the battery any more than is necessary. The charge controller will prevent the battery from overcharging and discharging.

The size of the battery bank can be calculated quite simply: for domestic installations the battery capacity should be at least five times your daily usage.

Let us assume that we have dedicated circuits with the following power requirements.

Lighting	300 W
Computer	200 W
Fridge	200 W
Freezer	200 W
TV	100 W
Boiler and pump	150 W

Now we need to look at how long these appliances will run for in a 24-hour period.

If we say that the lighting, TV and computer are going to be used for 6 hours a day, the energy required to run this part of the installation would be:

$$\text{Required energy} \times \text{Time in hours} = \text{Watt hours per day (Wh/day)}.$$

$$550 \times 16 = 8800\,\text{Wh/day}.$$

The fridge, freezer and boiler can be calculated for the worst case which may be 16 hours.

$$550 \times 16 = 8800\,\text{Wh/day}.$$

This gives us a total of $3600 + 8800 = 12{,}400\,\text{Wh/day}$.

We now need to calculate the requirement for 5 days:

$$5 \times 12{,}400 = 62{,}000\,\text{Wh}.$$

From this, we can calculate the AH requirements for the battery bank. We must remember that there will be some energy losses in the system. These systems are expected to operate with an efficiency of 80%.

The battery bank will need to have a capacity of:

$$\frac{\text{Total load} \times \text{Days of storage}}{\text{System voltage} \times \text{Efficiency}} = \text{AH of storage}$$

$$\frac{62{,}000}{12 \times 0.8} = 6458.33\,\text{AH}$$

If we say our batteries are 270 AH we will need:

$$\frac{6458.33}{270} = 23.91\ \text{batteries}$$

The calculation shown is for guidance; with careful consideration it would be possible to reduce the size of the battery bank. For instance, low-energy lighting could be used, and it may be that your fridge and freezer do not actually run for 16 hours. Each installation must be treaded individually and designed accordingly.

Voltage drop

It is very important to keep the voltage drop as low as possible on the d.c. side of the installation, since loss of voltage will have a big impact on power losses which in an efficient system must be kept to a minimum. Although BS 7671 allows a voltage drop of 5% for all circuits other than lighting circuits, document G83/1 states that 3% is the maximum for PV arrays. It is far better to work on a value near to 1% where possible.

If we use the values from the example which we have used previously we can see that string 1 has an I_{stc} of 7.7 A and a V_{oc} of 154 V. If we are to wire each string back to the inverter separately we would need to ensure that the correct cable size was used in relation to current and voltage drop. We can calculate this two ways: we can calculate the length permissible for a certain cable, or if we know the length we can simply calculate the minimum size required for a 1% drop.

A 1% drop can be calculated:

$$\frac{154 \times 1\%}{100} = 1.54\,\text{V}$$

To find out the maximum length permissible for a 2.5-mm^2 cable under these conditions:

From table 4F3B we can see that for a 2.5-mm^2 flexible cord the mV/A/m is 19. The calculation is:

$$\frac{\text{Voltage drop} \times 1000}{\text{Millivolts} \times \text{A}} = \text{Length}$$

$$\frac{1.54 \times 1000}{19 \times 7.7} = 10.52\,\text{m}$$

If our run is no longer than 10.52 m, a 2.5-mm^2 cable may be used; we must remember though that the length of cable must include the connection of the PV panels.

A far better method would be to measure the distance and then work the size of the cable out; this can be done as follows.

Let us assume that we need to use a cable 20 m long. The calculation for this is:

$$\frac{\dfrac{\text{Voltage drop} \times 1000}{\text{Length} \times \text{A}} = \dfrac{\text{mV}}{\text{A}}}{\text{m}}$$

$$\frac{\dfrac{1.54 \times 1000}{20 \times 7.7} = 10\dfrac{\text{mV}}{\text{A}}}{\text{m}}$$

As table 4F3B only gives sizes up to 4 mm^2, we now need to refer to table 4F1B, which will provide us with voltage drop values for cables above 4 mm^2.

We need to select a cable with a mV/A/m value of less than 10, and we can see that we will need to use 6 mm^2.

It must be remembered though that the 1% voltage drop is not a requirement, it is being used simply to keep power losses to a minimum. The voltage which we used was the open circuit voltage, which is not going to be the voltage under load;

however, it is always better to work with the worst case. If we were to choose to use a 4-mm^2 cable then the voltage drop would be:

$$\frac{\text{mV} \times \text{A} \times 1}{1000} = \text{Voltage drop}$$

$$\frac{12 \times 7.7 \times 20}{1000} = 1.84\,\text{V}$$

This is only 1.2% so the losses would be acceptable.

The use of the correct connectors is also very important with regard to keeping losses to a minimum.

Earthing and bonding

Equipment used in solar PV installations is usually Class II equipment and will not require earthing or bonding; however, if Class I equipment is to be used, then the same rules would apply as for a standard electrical installation and it would need earthing and bonding.

Summary

PV arrays consisting of not more than three strings do not require overcurrent protection; however, the cable must be capable of carrying 1.25 times the I_{sc} (short circuit current). Where more than three strings are used, the cables must be protected by fuses in the positive and negative conductors. The string fuse must have a tripping current of less than $2I_{sc}$ and the cable current carrying capacity, whichever is the smaller.

Each string must have a d.c. isolator and it is preferable to have an isolator as near to the string as possible and another at the inverter or controller (Figure 7.14).

The a.c. side of the inverter must also have a dedicated isolator which effectively isolates the PV installation from the supply.

Where the equipment used for the PV installation is Class II equipment, earthing and bonding are not required.

Always use purpose-designed PV cables for a PV installation (HO7RNF cable). Remember the PV array can get hot and any cable which is routed behind the array must be rated for a minimum temperature of 80°C. The cable should also be marked (Danger PV array cable-high voltage d.c. during daylight).

All connecters and switches on the d.c. side of the installation must be d.c. rated. For switching inductive loads a.c. switches are suitable.

It is better to have dedicated circuits for the PV installation. If this is impractical, then it is preferable for the PV supply from the inverter to be connected to the electrical installation via its own dedicated consumer unit. In all cases, it must be connected to the installation via a protective device; for this reason it makes sense to combine the a.c. inverter isolation and the protective device in one unit. This will also alleviate the need for a periodic inspection on the existing installation.

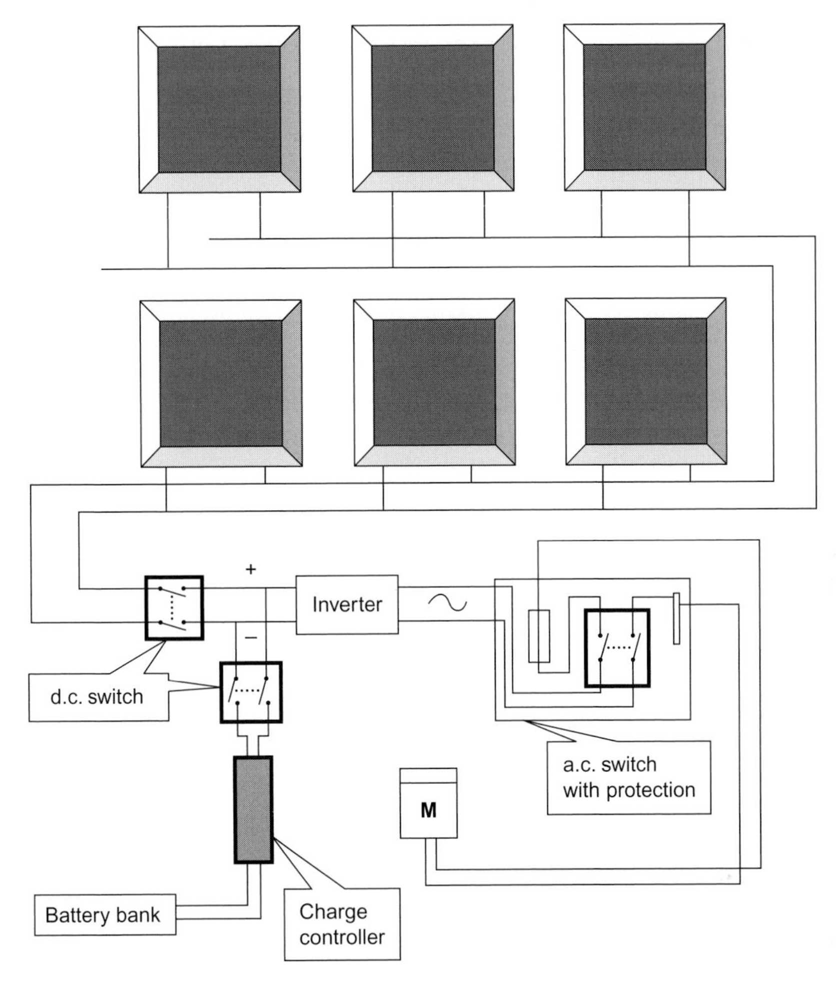

FIGURE 7.14 **Grid connected with battery backup**

SECTION 717

Mobile or transportable units

These units are generally used for outside events such as filming, catering, accident scenes or any other occasion where a mobile or transportable unit may require an electrical supply to enable it to operate. The supply to the vehicle can be from an outside source which is plugged into the vehicle or can be from a self-contained low-voltage generating set. The units can be either self-propelled or on a trailer.

Mobile and transportable units are classed as special installations due to their being mobile; this in itself could cause problems due to vibration, loss of connection to earth due to the use of temporary cable connections, and high protective conductor currents, particularly where the unit is being used for communications. Another risk which is common to many special installations is the lack of an equipotential zone.

The electrical connection to a unit can be by a TN-C-S, TN-S or TT system; however, where a TN-C-S system is used, it must be continuously under the supervision of a skilled or instructed person, and the connection to earth must be verified before the unit is connected. In most cases, it is better to convert the supply to a TT system at the unit.

The connection to the unit must be by an industrial type plug and socket to BS EN 60309-2 with a minimum rating of IP44. The plug part of the socket outlet must be part of the unit.

Due to the problems associated with this type of installation there is a seven-point checklist which must be completed:

1. Check the suitability of the electrical supply before connecting the unit.
2. Use RCDs.
3. All accessible conductive parts of the unit must be bonded to the main earth terminal of the installation.
4. Regularly inspect and test all cables and couplers; this must be recorded.
5. Use flexible cables.
6. Use clear labelling indicating which type of supply may be connected to the unit.
7. The unit must be maintained and periodically inspected; this should be documented.

As well as this checklist there are other requirements which will enhance the safety of these units.

On completion of the installation an initial verification must be carried out on both the d.c. and a.c. sides of the installation. An electrical installation certificate must be completed for the a.c. part of the installation and a commissioning test sheet must be completed for the d.c. side of the installation.

Socket outlets used to provide a supply for use outside the unit should preferably be of the industrial type and have an IP rating of at least IP44; these socket outlets as well as those within the unit must be protected by an RCD with a maximum rating of 30 mA. They may also be SELV, PELV or supplied by an isolating transformer; in these instances RCD protection is not required.

A notice must be positioned near the electrical inlet to the unit. This notice must indicate:

- The types of supply which can be connected to the unit
- The number of phases and the voltage rating
- The power demand of the unit
- The earthing arrangement for the unit.

Often these types of unit have cooking facilities which require the use of gas cylinders. Electrical equipment must not be installed in a compartment which is used to store gas cylinders; however, it is permissible to allow cables to pass through the compartment provided the cables are protected against mechanical damage and are installed in a gas-tight conduit (it only needs to be gas tight where it passes through the compartment).

Most of these units will require the use of an earth electrode connected to the main earthing terminal; this will ensure that any exposed and extraneous conductive parts are kept at the same potential as earth.

SECTION 721

Electrical installations in caravans and motor caravans

This section only applies to installations in caravans and motor caravans which are used for recreational purposes. It does not apply to static caravans which are used to live in.

Caravans are classed as special locations for many of the same reasons that caravan parks, marinas and mobile or transportable units are. It is virtually impossible to establish an equipotential zone around them; there is likely to be a lot of vibration, and they are connected to a supply by the use of a flexible cord.

All exposed and extraneous conductive parts that can be touched from within the caravan must be bonded to the main earthing terminal.

Protection must be by ADS, which includes a double pole RCD with a maximum rating of 30 mA and an overcurrent protective device which is also double pole. It must also be possible to isolate the electrical supply by the use of one switch; this can be the RCD provided it is accessible from within the caravan. Where the installation consists of only one circuit, the double pole circuit breaker can be used for isolation.

The wiring can be stranded singles installed in plastic conduit or twin and earth, whichever is the most suitable.

Electrical equipment must not be installed in a compartment which is used to store gas cylinders; however, it is permissible to allow cables to pass through the compartment provided the cables are protected against mechanical damage and are installed in a gas-tight conduit (it only needs to be gas tight where it passes through the compartment).

The supply can be connected by the use of an industrial type appliance inlet complying with BS EN 60309-1. The inlet must be located in an accessible position no higher than 1.8 m from ground level; it must also comply with a minimum of IP44. The cable used for connecting the caravan must not be any longer than 27 m and have a minimum cross-sectional area as shown in Table 7.4.

TABLE 7.4 Conductor sizes

Rated current (A)	Cross-sectional area
16	2.5 mm^2
25	4 mm^2
32	6 mm^2
63	16 mm^2
100	35 mm^2

A notice must also be given with instructions for use of the caravan; it must include a description of the installation, the position and operation of the main switch and the use of the RCD test buttons. There is also a set of instructions for the electricity supply contained in section 721 of BS 7671 which must be copied and installed within the caravan.

Wherever possible, luminaires should be fixed directly to the caravan structure. Where this is not possible or the use of a pendant is preferred, a method of securing the pendant must be provided to prevent it swinging around and causing damage when the caravan is moving.

Where the caravan uses a supply from a car battery or other extra-low-voltage source annex A at the end of section 721 provides some very useful information; this includes the wiring of the plug and socket outlets which are required for towing purposes.

SECTION 740

Temporary electrical installations for structures, amusement devices and booths at fairgrounds, amusement parks and circuses

This section of the regulations is really aimed at travelling circuses and fairs; this is an area in which most electricians would never be involved. This type of installation is one which is moved repeatedly and is usually connected by the showmen – in circuses this could be the clown, ringmaster or trapeze artist. I could read through the text in BS 7671 and provide a safe electrical installation, or even describe how to install a safe installation in a big top if required. However, as most of us will never get involved with this type of work I feel that it would be of little interest. I did try and look around one of the large generators at a fairground once, but unfortunately I was put off by a rather large dog.

SECTION 753

Floor and ceiling heating systems

The reason that these types of systems become special locations is that they are concealed within the structure. Usually, once they have been installed only the controls are accessible.

Due to the elements being concealed, there is a risk of penetration of the heating elements as well as a fire risk to the actual building structure.

All circuits supplying floor and ceiling heating systems must be protected by an RCD with a maximum rating of 30 mA.

As with any other part of an electrical installation heating elements must comply with the relevant British Standards, and they must also be installed as instructed by the manufacturer.

Great care must be taken to ensure that the heating elements are not damaged during or after installation; the elements must not be subjected to movement after installation. For that reason, they must not be installed across an expansion joint, or any other part of a structure which may move.

Heat-free areas must also be provided to allow the fixing of other types of equipment where required; it is a good idea to provide a drawing with measurements showing where the heat-free areas are. The most common use of underfloor electric heating is within bathrooms and shower rooms. Before the heating is installed, it is very important that the location of bathroom suites and kitchen furniture is known, as any part of the heating installation which passes under any installed equipment will overheat; this will present a fire risk or overheat and damage the heating element. All areas of the floor or ceiling which contain this type of heating must be open to free air to allow safe dissipation of heat.

Where the heating units are connected to the cold tails, the connections must be of a permanent nature. The simplest method is to use crimps, but it must be remembered that although they are called cold tails they may still become hot due to the transmitted heat from the heating elements.

The zone which is being heated must be limited to a maximum temperature of 80°C with the floor area being limited to 35°C where contact with bare skin or footwear is likely to occur. This can be achieved by correct design and installation. Wherever there is any doubt as to the design or suitability of a floor or ceiling heating system the manufacturer will always provide you with the best advice and should always be contacted.

APPENDIX 1

Appendix 1 is simply a list of British or BS EN standards and where they can be referenced within BS 7671. The first section contains reference to British Standards only; the second section is a list of the British Standards which have been accepted as the European Norm (BS EN). The appendix only makes reference to them and the complete documents can be purchased from the British Standards Institute. BS 5266 for emergency lighting and BS 5839 for fire detection and alarm systems in building are the ones which are most commonly required by electricians.

APPENDIX 2

A list of statutory regulations which may have to be applied in some installations is provided here along with some other bits of information which will help an electrician/designer reference the correct documentation for certain types of electrical installation.

The last part of appendix 2 shows a table of the permitted voltage drops in a supply.

A 230-V supply has a tolerance of +10% (23 V) and −6% (13.8 V). This will give a maximum of 253 V and a minimum of 216.2 V.

As we can see from appendix 12, a final circuit is permitted to have a maximum voltage drop of 6.9 V for lighting, and 11.5 V for all other circuits.

This means that in some cases where the supply is at the minimum voltage of 216.2 V, and the cable calculation has used a design voltage of 230 V, the voltage at the load end of the cable could be 216.2–11.5 = 204.7 V. Some equipment may struggle to run at this low voltage and various problems can occur; it may be better to measure the supply voltage before carrying out voltage drop calculations and then to apply the percentages permitted to the actual voltage rather than 230 V, which should always be taken as a maximum for single-phase calculations. Where the supply and the load are three phase then the percentages are applied to 400 V, again as a maximum.

APPENDIX 3

This appendix provides us with information on the time and current characteristics of overcurrent protective devices and RCDs.

The first calculation is $Z_s = U_o / I_a$.

A Practical Guide to the 17th Edition of the Wiring Regulations. DOI: 10.1016/B978-0-08-096560-4.00012-6

Z_s is the maximum earth fault loop impedance at which the protective device could be relied upon to operate.

U_o is the nominal line voltage (a.c.) to earth (taken as 230 V in the majority of cases, no need to measure). Obviously, where different voltages are used such as 110 V, these values are used (55 *V is* U_o *for a* 110-*V construction transformer. See chapter* 41 *table* 41.6).

I_a is the current which will cause the protective device to operate within the specified time.

The current curves in appendix 3 show the worst case current (highest). Manufacturers of protective devices provide tables showing the lowest and highest currents. This is because the devices have to operate within a current window, and this can vary depending on the type of device. The current windows for BS EN 60898 devices are shown in Table A1.

We use as an example a 16-A type C circuit breaker with a supply voltage of 230 V (U_o).

The circuit breaker will require between 5 and 10 times its rating to ensure operation within 0.1 s. We must use the worst case; if the circuit breaker operates at less current it is fine as long as it is not less than 5 times its rating.

The worst case is:

$$10 \times 16 = 160\,\text{A}$$

The 16 A circuit breaker will require 160 A to ensure that it operates in the correct time (0.1 s).

Using the formula given in appendix 3, we can see that the maximum Z_s permissible would be 230/160 = 1.44 Ω (rounded up).

If we now look at figure 3.5 for type C circuit breakers, we can see that at a fault current of 160 A the circuit breaker will operate at 0.1 s; the table in the top right corner shows that the current required to operate the circuit breaker is 160 A. If we

TABLE A1 Circuit breaker trip current and uses

Type of circuit breaker	Minimum and maximum trip current	Type of use
Type B	3 to 5 times I_n	General purpose with no inrush currents due to switching
Type C	5 to 10 times I_n	For use where small inrush currents may be encountered such as a number of fluorescent lights or a small motor
Type D	10 to 20 times I_n	Must be used with care; most suitable for circuits with high inrush currents such as large transformers and X-ray machines

look in table 41.3, we can see that the maximum permitted Z_s for a 16-A type C circuit breaker is 1.44 Ω.

All circuit breakers, if operating correctly, will operate within 0.1 s provided the required current is allowed to pass through them. Fuses are entirely different and these can be installed for different disconnection times; these times can be up to a maximum of 5 s for some circuits or a maximum of 0.2 s for other circuits. Table 41.1 provides information with regard to disconnection times.

A 230-V circuit up to 32 A supplied by a TN-S or TN-C-S supply must have a disconnection time of 0.4 s; if it is supplied by a TT system it must have a disconnection time of 0.2 s.

Any circuit over 32 A or a distribution circuit must have a disconnection time of 5 s maximum. Where the distribution circuit is fed from a TT system, the disconnection time will be a maximum of 1 s.

Remember, where circuit breakers are used and installed correctly, they will operate at a maximum of 0.1 s. Consideration need not be given to 5 s, 0.4 s and 0.2 s disconnection times.

We can use these tables for calculating the actual disconnection time for a given circuit; this can be very useful, particularly where fuses are being used.

Example

Let us assume that we are calculating a circuit which must disconnect in the event of a fault in 0.4 s. The protective device is a 20-A BS 88 fuse. The information that we have is that the Z_e is 0.35 Ω and that the circuit $R_1 + R_2$ is 1.23 Ω.

First, we must calculate the $R_1 + R_2$ for its operating temperature of 70°C by multiplying it by 1.2.

$$12 \times 1.13 = 1.356\Omega$$

Now we can calculate Z_s.

$$Z_s = 0.35 + 1.356 = 1406\Omega \quad Z_s = 1.706\Omega$$

Now we must calculate the current which will flow in the event of a fault (I_f).

$$I_f = \frac{U_o}{Z_s}$$

$$I_f = \frac{230}{1.706} = 134\,A$$

If we now look at figure 3.3A in appendix 3, we can see that the device will operate at very slightly less than 0.4 s.

Appendix 3 also provides information on the required performance criteria for the correct operation for RCDs to BS EN 61008-1 and 61009-1. Figure A1 provides information for these RCDs and also BS 4293.

Standard	Type	$I\Delta_n$	$\frac{1}{2} \times I\Delta_n$ (t)	$I\Delta_n$ (t)	$5 \times I\Delta_n$
BS EN 61008	G	≤30 mA	≥2000 ms	≤300 ms	≤40 ms
	S	≤30 mA		130–500 ms	N/A
BS EN 61009	G	≤30 mA	≥2000 ms	≤300 ms	≤40 ms
	S	≤30 mA		130–500 ms	N/A
BS 4293	G	≤30 mA	≥2000 ms	≤200 ms	≤40 ms
	Time delay	≥30 mA		300–400 ms	N/A

FIGURE A1

APPENDIX 4

All electrical installations have current-carrying conductors of some kind; usually these conductors form part of a cable or flexible cord, and occasionally the conductors are bus bars contained in trunking. This section of BS 7671 deals with the current-carrying capacity and voltage drop for cables and flexible cords.

For the installation to be safe, the conductors have to be the correct size, and be capable of operating at the load which they have been designed to carry continuously without damage. This involves selecting a cable with the correct type of insulation capable of withstanding the temperature in which it is to be installed. The cable must also be selected with the correct type of mechanical protection suitable for the environment.

Ambient temperature

Let us look at temperature first, as this really is an electrician's biggest enemy. It is important that the cable never exceeds its recommended operating temperature; as we have seen earlier in this book most of our cables will be required to operate at 70°C, due to the fact that most of our accessories are rated at this temperature.

In some special instances cables may be installed at higher temperatures; there is no problem with this provided the cable and the equipment are suitable for the temperature involved and the cable is installed using a suitable installation method.

As most of our installations will involve a cable operating at 70°C this is what we will use here. The cable current-carrying capacity tables in appendix 4 assume that our cables are installed in an ambient temperature of 30°C, and in most cases provide us with the maximum current which the conductor can carry before the temperature of the conductor will exceed 70°C. If we use table 4D1A, we can see that a 2.5-mm^2

single-phase cable enclosed in conduit (*column* 4) can carry 24A (I_t); this is provided in the first instance that the ambient temperature is 30°C. In basic terms, if we pass a load of 24A through a cable which is installed at 30°C, the temperature will rise to 70°C.

Where the ambient temperature is different from 30°C, an adjustment has to be made to compensate for the different temperature. This adjustment is made using rating factors for ambient temperature, which can be found in table 4B1.

Example 1

Let us say a 2.5-mm^2 cable is to be installed in an area which has an ambient temperature of 40°C. We can see from table 4B1 that a 70°C thermoplastic cable has to have a rating factor of 0.87(C_a) applied to it; this will give us the maximum current that the cable could carry without the conductor temperature rising above 70°C: $24 \times 0.87 = 20.88$A. This value is known as I_Z (*I_Z is the current which a cable can carry under its installation conditions*).

This shows us that due to the cable starting at a higher temperature it will require less current to take its temperature to 70°C.

Grouping

Another problem which affects our cables is temperature which is passed to them from other cables; if we install a cable into an enclosure such as a conduit or trunking which contains other cables already having a conductor temperature of 70°C, it stands to reason that some of the temperature will be passed from one cable to another. This also has to be taken into account when cables are installed alongside cables from other circuits; for this we use a rating factor for grouping (C_g) and these can be found in table 4C1.

Example 2

Our circuit is installed in a conduit which contains two other circuits. If we look at table 4C1 we see that a rating factor (C_g) of 0.7 must be used because the conduit now contains three circuits. Using column 2 from table 4D1A, we can see that a 2.5-mm^2 cable has a current-carrying capacity (I_t) of 24A.

I_Z will now be $24 \times 0.7 = 16.8$A.

Due to being enclosed with other circuits, a 2.5-mm^2 cable can now only carry 18.9A. This is assuming that all of our cables are fully loaded and operating at 70°C.

Very often our cables are not fully loaded; provided that they are operating at no more than 30% of their grouped rating factor they can be discounted.

Example 3

From the earlier examples, a 2.5-mm^2 cable can carry 24A before any correction factors are applied. To calculate the amount of current which the cable can carry before it needs to be counted as a circuit in our grouping calculation the following formula can be used:

$$24 \times 0.3 \times 0.7 = 5.04\,A$$

24 is the rating factor of the cable from column 4 of table 4D1A, 0.3 is the multiplier for 30% and 0.7 is the grouping factor used in Example 2.

Using this calculation shows us that if the 2.5-mm^2 cable was only carrying 5.04A, we could discount it from our grouping calculation; for the remaining circuits only a grouping factor for two circuits (0.8) need be applied.

The important thing to remember is that for a cable to be discounted for grouping it must only be carrying 30% of its capacity with the grouping factor for all of the circuits applied. Hence, in our example the grouping factor for three circuits was applied to the 2.5-mm^2 cable, which has been discounted in the calculation.

If we now look at Examples 1 and 2 again, it is possible that the cable could be installed in a conduit with two other circuits and the ambient temperature of the area where the conduit is installed is 40°C. In this case, both correction factors must be applied to the circuit to be installed; this would result in a further reduction of current-carrying capacity for the circuit.

Example 4

Ambient temperature factor (C_a) = 0.87
Grouping factor (C_g) = 0.7
Tabulated current rating of the cable is 24A

As the circuit is in conduit at 40°C and grouped with two other circuits both rating factors must be applied.

$$24 \times 0.87 \times 0.7 = 14.61A$$

As you can see, this now reduces the current-carrying capacity to 14.61A; this is due to the likely build-up of temperature in the cable from being at a greater ambient temperature than 30°C and the imported heat from the other circuits.

Where all the rating factors do not affect the cables at the same time only the worst case needs to be applied – in this case C_g 0.7. In other words, if the conduit passes through an environment at 40°C with only one circuit in and then enters another room which has a temperature of 30°C but picks up another two circuits in that room, only the worst case needs to be applied to our cable – this would be C_g 0.7.

In some installations, three- and single-phase circuits will be contained in the same enclosure. Table 4C1 only provides factors for uniform groups of cables; if all of the circuits are three phase then each three-phase circuit is counted as one circuit. The factors from the table can be used just as they can for single-phase circuits. This is not the case where there are single- and three-phase circuits mixed together, and a slightly different approach to grouping has to be taken.

To estimate the grouping factors for three-phase circuits in a containment system having mixed groups we must multiply the number of three-phase circuits by 1.5, then add this number to the number of single circuits. This will give us a figure to use with table 4C1.

Estimating the grouping factor for single-phase circuits in a mixed containment system, we must multiply the number of single-phase circuits by 0.666 and then add this to the number of three-phase circuits.

Example 5

A trunking contains 5 three-phase circuits and 3 single-phase circuits.

Equivalent number of single-phase circuits:

$5 \times 1.5 = 7.5$

$7.5 + 3 = 10.5$; in this case a rating factor of 0.51 could be used (4C1).

Equivalent number of three-phase circuits:

$3 \times 0.666 = 1.998$; round this up to 2.

$2 + 5 = 7$; in this case a rating factor of 0.54 could be used (4C1).

When calculating the cable size use the three-phase factor for three-phase circuits (*in this case* 0.54) and single-phase factor for single-phase circuits (*in this case* 0.51).

Thermal insulation

Another problem with heat in cables is where the circuits pass through thermal insulation. This subject often causes confusion, with the end result being that the cable ends up have rating factors for thermal insulation being applied twice.

Table 52.2 provides rating factors for use where cables are totally surrounded by thermal insulation. Where a cable passes through thermal insulation for a distance of 500 mm the cable rating must be halved.

If we look at table 4D5 in BS 7671, we can see that a 2.5-mm^2 cable can carry a maximum of 27 A if clipped direct (*method C column* 6); if we take the same cable and surround it with thermal insulation (*method* 103 *column* 5), we can see that its current rating is reduced to 13.5 A.

The calculation for this is the same as the previous examples; however, in this case the calculation is completed for us by using table 4D5.

If we take the value of 27 A for the cable clipped direct and apply our correction factor C_i, which in this case is 0.5, we can see how it works.

$I_Z = 27 \times 0.5$ which is 13 A.

Table 4D5 is specifically for 70°C twin and earth cable and each reference method takes into account the different installation methods, which are well explained at the top of each column. If a picture of these methods is useful then table 4A2 should be used.

Table 4D5 takes into account thermal insulation and where this table is used no other rating factor for insulation should be applied.

Where other cables which may come into direct contact with thermal insulation are used then the rating factors from table 52.2 must be used; however, where the cables are in a containment system such as conduit or trunking, the correct use of the installation method from table 4A2 will compensate for thermal insulation.

Example 6

Let us take a conduit which has been chased into a wall built of thermal blocks; the conduit contains a single-phase circuit which is wired in single-core 70°C thermoplastic insulated cables which are required to carry a load of 22 A.

The first step is to identify the installation method by using table 4A2. To be accurate with this, we need to know the thermal resistivity of the wall in which the conduit is installed; as with all other calculations if this is not known and cannot be found by enquiry, we must use the worst option.

Most blocks used in modern building will not have a thermal resistivity of 2 km/W; if we refer to table 4A2 we can see that number 59 is for use with non-sheathed cables in conduit in a masonry wall having a thermal resistivity of not greater than 2 km/W. Reference method B is to be used in this case.

We now need to refer to table 4D1A which is for use with the cables we are installing. If we look along the headings at the top of the page we will see reference method B; just below that is column 4 which is for two cables, single phase. A 2.5-mm^2 cable can carry 24 A under these conditions.

As you can see, thermal insulation is accounted for in these tables; it is only where the cables are completely surrounded by thermal insulation that table 52.2 needs to be used.

The section in this book on overload describes the reason for using a rating factor of 0.725 when a rewirable fuse is used for overload protection; C_c is the symbol for this factor. Where we use this factor, it must be used as a divisor into the fuse rating and not as a multiplier of the cable rating.

Where we use the rating factors C_a, C_g and C_i as multipliers to the cable rating we are calculating the amount of current which could be carried by the cable under the conditions imposed on it.

Example 7

We have a 2.5-mm^2 twin and earth thermoplastic 70°C cable installed in conduit with two other circuits; the conduit is clipped direct to a masonry wall but it passes through 100 mm of thermal insulation at an ambient temperature of 35°C. It is protected by a 20-A BS EN 60898 circuit breaker.

We now need to reference the rating factors. Grouping (C_g) is the first; this can be found in table 4C1: three circuits enclosed is 0.7.

Next is thermal insulation (C_i), which is found in table 52.2 and is 0.78, and finally we need C_a ambient temperature, which is found in table 4B1 and is 0.94.

Our correction factors are:

C_g	0.7
C_i	0.78
C_a	0.94

From the example we are using, we know that the minimum current rating of the cable has to be 20 A, as this is the rating of the protective device and we are protecting the circuit from overload. As it is enclosed in conduit but clipped direct we can use method C; we can use this method because we are using separate factors for insulation.

To calculate the current which a cable can carry under these conditions (I_Z), we need to select a cable; let us try 4 mm from column 6.

$$37 \times 0.7 \times 0.78 \times 0.94 = 18.98\,A$$

Under the conditions given, a 4-mm^2 conductor can only carry 18.98 A. As we have a 20-A protective device this cable will be unsuitable; we will need to try 6 mm^2 which clipped direct will carry 47 A before any factors are applied.

$$47 \times 0.7 \times 0.78 \times 0.94 = 24.12\,A$$

I_Z is 24.12A, so this cable will be fine.

This approach to calculating the correct size of cable is acceptable, but often it requires a couple of calculations. A simpler approach is to use all of the factors as divisors, particularly where a rewirable fuse is used, as the factor for this (*always* 0.725) has to be used. To use the factors as divisors:

$$I_t = \frac{I_n}{(C_a \times C_i \times C_g)}$$

$$I_t \geq \frac{20}{(0.94 \times 0.78 \times 0.7)} = 38.96\,\text{A}$$

We now need to select from column 6 of table 4D5 a cable which is capable of carrying 38.96A, which as we can see is 4mm². We have ended up with the same cable size but have only carried out one calculation.

If the protective device was a rewirable fuse to BS 3036 we would probably need a larger cable:

$$I_t \geq \frac{20}{(0.94 \times 0.78 \times 0.7 \times 0.725)} = 53.74\,\text{A}$$

This would require a 10-mm² cable. It is usually better to use a circuit breaker or a cartridge fuse than a BS 3036 rewirable fuse, and also with a little care we can often eliminate the other correction factors. Wherever we can we should avoid grouping and thermal insulation; it is often better to take a longer route for the cable if it means we can use a smaller cable.

We have to be careful, however, because we also have voltage drop to consider.

Voltage drop

Appendix 12 provides us with information on voltage drop; table 12A shows that we are allowed a maximum of 3% (6.9V) for lighting and 5% (11.5V) for all other circuits.

To calculate voltage drop we need to refer to the tables in appendix 4 again.

Let us say we are going to use an armoured 70°C thermoplastic cable which is to be installed on to a cable tray; it is a 10-mm² cable and it needs to carry a single-phase load of 65A for a distance of 53m. If we refer to table 4D4A, we can see that the cable can carry 72A: so far, so good! Now we need to look at table 4D4B to find out the voltage drop per metre; we can see from column 3 that a 10-mm² cable will have a voltage drop of 4.4mV/A/m.

$$mV = 4.4$$

$$A = 65$$

$$m = 53$$

$$v/d = \frac{4.4 \times 65 \times 53}{1000} = 15.15\,\text{V}$$

The 1000 *is used in the formula because the value from our tables is in millivolts and we need to convert it to volts.*

Clearly, if we are only allowed a maximum of 11.5 V this cable will not be suitable; we could carry out the same calculation for a larger cable. However, it would have been better to use a different method and calculate the maximum mV/A/m for a cable, and then choose it by voltage drop rather than size.

We have to transpose the original formula:

$$\frac{mV \times A \times m}{1000} = \frac{v}{d}$$

$$mV = \frac{vd \times 1000}{A \times m}$$

We know that the maximum voltage drop permissible is 11.5 V, the current in the cable is 65 A and the length of run is 53 m. If we enter these values into the formula:

$$\frac{11.5 \times 1000}{(65 \times 53)} = 3.33\,\text{mV}$$

We now need to look in table 4D4B for a conductor which has a voltage drop of 3.33 m/V/m or less and we can see that 16 mm^2 will be fine.

You will notice that for all cables over 16 mm^2 there are three values given for voltage drop, *r*, *x* and *z*. There is a complex calculation in which these values can be used to calculate a reduced voltage drop within a cable depending on the conductor temperature.

The object of this book is to try and simplify the regulations as much as possible; my advice to anyone using cables greater than 16 mm^2 would be to use the value *r* in the voltage drop calculation as this will always give a value which is on the right side of safe, or to use a cable calculation programme.

Voltage drop in ring circuits

Where the load at each point of the ring circuit is known, then an accurate calculation of voltage drop can be achieved. In most cases this is not possible due to the nature of the use of a ring circuit.

Where the length of the ring circuit is known, then the calculation is quite simple.

Example 8

A ring circuit is wired in 2.5 mm^2 70°C twin and earth thermoplastic copper cables; the circuit is 72 m from end to end and is protected by a 32-A circuit breaker.

If we look in table 4D5A column 8, we can see that the voltage drop for 2.5 mm^2 is 18 mV/A/m. The voltage drop over the length of the cable will be (18 × 32 × 72) / 1000 = 41.47 V.

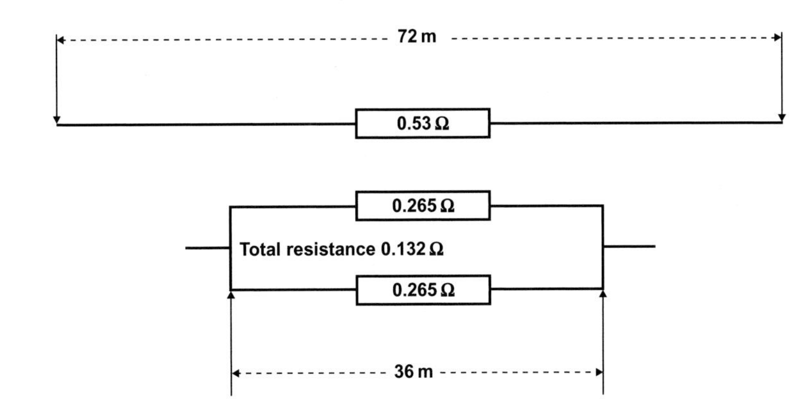

FIGURE A2

This looks bad but we must remember that the cable is in the form of a ring, which means that the cable will be half the length and twice the cross-sectional area.

We must now divide the voltage drop by 4, which will give us the true voltage drop if the circuit was fully loaded to 32 A.

$$\frac{41.47}{4} = 10.36\,\text{V}$$

This is a pessimistically high value and where the load is known to be less, the value of the load current can be used instead of the rating of the protective device. This will allow a longer circuit to be installed. A 2.5-mm^2 conductor has a resistance of 7.41 mΩ/m; 72 m of this conductor will have a resistance of (Figure A2):

$$\frac{72 \times 7.41}{1000} = 0.53\,\Omega$$

Note: Two 36 *m lengths of* 2.5 *mm*^{2} *cable in parallel will have a resistance of* 0.53/4 = 0.132 Ω *(*R_1*).*

0.132 Ω *will be the resistance of the line conductor, but the current flows in the neutral conductor as well. Therefore, the resistance of the ring must include this resistance as well (*R_n*), so the total resistance of the ring will be* 0.264 Ω.

We must also remember that this resistance will be at 20°C *and that the conductors may be at* 70°C *when they are carrying current. The correction for temperature is* 1.2: $0.264 \times 1.2 = 0.317\,\Omega$.

If we now use Ohm's law, the maximum current which may flow in the circuit is $I \times R$ *or* $32 \times 0.317 = 10.144$ V.

This is the voltage drop in the circuit; it is slightly lower than the value calculated using the voltage drop given in the tables in appendix 4, *because the* 18 *mV/A/m is rounded up.*

Example 9

A ring circuit is to be wired in 2.5-mm² twin and earth cable with a total known load of 23 A.

We can calculate the maximum length of circuit by transposing the formula used in Example 8. (*The 4 used in the calculation is because the ring is half the distance and twice the CSA.*)

$$\frac{mV \times A \times m}{(1000 \times 4)} = vd$$

$$\frac{vd \times 1000 \times 4}{(mV \times A)} = m$$

If we now enter the known values:

$$\frac{11.5 \times 1000 \times 4}{(18 \times 23)} = 111\text{m}$$

The maximum length of the circuit carrying 23 A can be 23 m.

Cable calculation

We can now put all of the information which we have gained to use by carrying out a cable calculation.

A multicore 90°C armoured thermosetting cable is to be used to supply a load of 4.8 kW. The load is 28 m from the origin of the circuit which is to be protected by a BS 88 cartridge fuse. The cable is to be clipped direct to a masonry surface alongside one other circuit at an ambient temperature of 30°C. Due to the nature of the load a maximum voltage drop of 5 V is permissible, the supply is 230 V a.c. with a Z_e of 0.24 Ω and the cable sheath is to be used as the CPC.

The process is:

$$\text{Design current } (I_b) = \frac{4.8}{230} \times 1000$$

$$I_b = 20.86\,\text{A}$$

Protective device size $I_n \geq I_b$ (*if you are unsure of the sizes available look in table* 41.2).

$$I_n = 25\,\text{A}$$

Rating factors

We have a rating factor for grouping to find, we must look in table 4C1. The rating factor for two circuits on a surface is 0.8.

$$C_g = 0.8$$

Ambient temperature is 30°C; from table 4B1 we can see that the rating factor is 1.

$$C_a = 1$$

These are the only factors which will affect our circuit in this instance. We can now calculate the minimum conductor size.

$$I_t \geq \frac{I_n}{C_a \times C_g}$$

$$I_t \geq \frac{25}{(0.8 \times 1)} = 31.25\text{ A}$$

$$I_t = 31.25\text{ A}$$

At this point we can look in the tables for current-carrying capacity in appendix 4. We must remember that although the cable we are using is 90°C thermosetting we cannot allow it to rise in temperature above 70°C. For that reason we must use table 4D4A for 70°C thermoplastic.

From column 2, the conductor size that can carry 31.25 A is 4 mm^2.

We must now check the voltage drop to ensure that it is not greater than the maximum permitted for this circuit, which is 5 V.

From table 4D4B column 3, we can see that the mV/A/m for a 4-mm^2 copper conductor is 11 mV/A/m.

Our calculation is:

$$vd = \frac{mV \times A \times m}{1000}$$

$$\frac{11 \times 20.86 \times 28}{1000} = 6.42\text{ V}$$

This is too high and we must select a larger cable; if you remember, we looked at an easier way of getting the correct cable with one calculation, by transposition.

$$\frac{5 \times 1000}{(20.86 \times 28)} = \text{Maximum millivolts per m}$$

$$= 8.56\text{ mV/A/m}$$

If we now look at table 4D4B, we can see that a 6-mm^2 cable will be suitable.

This proves that the cable can carry the current and that the voltage drop requirements will be met. As we saw earlier in this book, this calculation is only part of the process; we now need to calculate to see if the protective device will operate in the required time. We no longer need to use appendix 4 but we may as well carry on and complete the calculation.

For this we need BS 5467 or the manufacturer's data for 90°C steel wired armour cable as we need to ensure that the CSA of the steel armour is suitable to be used as a CPC.

Table 54.7 of BS 7671 provides us with formulae for calculating the minimum CSA of the steel wire armour.

$\frac{k_1}{k_2} \times S$ is the calculation to give us the minimum size SWA.

k_1 is the value given to the line conductor and in this case it can be found in table 43.1 or 54.3; the value will be the same whichever table you use (*do not forget we are using* 90°C *cable but must calculate it at* 70°C).

$$k_1 = 115$$

k_2 is the value given to the protective conductor and can be found in table 54.4.

$$k_2 = 51$$

Using these values we can see that the minimum size SWA permitted will be $(115/51) \times 6 = 13.5\,\text{mm}^2$.

From our British Standard values for SWA cable we can see that the actual CSA of the steel armour is $22\,\text{mm}^2$; as this is larger than the value required it will be fine to use; however, we now need to check the disconnection time and for that we will need a value of $R_1 + R_2$ for the cable.

For this we need to refer to the British Standard or manufacturer's data for the cable again and we can see that the resistance of the steel is $7\,\text{m}\Omega/\text{m}$ and that the resistance of the copper conductor is $3.08\,\text{m}\Omega/\text{m}$; this will make the $r_1 + r_2$ value for the cable $10.08\,\text{m}\Omega/\text{m}$.

The total resistance ($R_1 + R_2$) is: (*do not forget the* 1.2 *for temperature*)

$$\frac{10.08 \times 28 \times 1.2}{1000} = 0.338\,\Omega$$

$$R_1 + R_2 = 0.338\,\Omega$$

Z_s for the circuit is $Z_e + R_1 + R_2$

$$0.24 + 0.338 = 0.578\,\Omega$$

We now need to compare this value with the Z_s value for the 25-A BS 88 protective device. As the circuit is rated at below 32A, we need to look at table 41.2.

The maximum permitted Z_s for this circuit is 1.44 Ω; our calculated value is 0.578 Ω, which is less than the maximum permitted, and the circuit will comply with the requirements of BS 7671.

As the cross-sectional area of the SWA complies with table 51.7, there is no need to check for thermal constraints (adiabatic calculation).

APPENDIX 5

Classification of external influences

External influence is defined in part 2 as 'any influence external to an electrical installation which affects the design and safe operation of that installation.'

These external influences are shown by the use of a code which has two capital letters and a number.

The first letter shows the category of external influence, the second letter shows what the external influence relates to and the number indicates the level of the external influence: the higher the number the worse the influence.

The first letter indicates which category the external influence falls under.

A is Environment (*wet, dry, humid, dust, vibration, etc.*).

B is Utilisation (*How is the building being used? What is the capability of the persons using the building? Are they ordinary persons or perhaps are they handicapped? How difficult would evacuation be if required? Does the building contain materials which could explode?*).

C is Construction of buildings (*What is the building constructed from? Are the materials flammable? Will it be likely to move?*).

The second letter indicates what the external influence could be; for instance if we refer to appendix 5 and look in the environment section (category A) we can see that if the second letter was K it would refer to flora (plants, moss and the like).

The last digit is a number and it indicates the level of the problem which the external influence is likely to be.

As an example if we look at an external influence given a code of AD1, we can see that the letter A relates to the environment, the letter D relates to water and the final digit 1 tells us that the problem likely to be caused by water is negligible.

When we require further information with regard to this external influence, we need to look further into appendix 5. When we find AD1 we can see that it relates to areas where weather protection is not required and that the IP rating for the equipment is IPX0. As the external influence relates to water nothing is specified for dust (X) and the level of protection against water is 0 which means no protection required. At the other end of the scale, we can see that AD8 is where the equipment is likely to be totally submersed, such as in a swimming pool. The IP rating given as protection is IPX8: X shows that nothing is specified for dust and 8 shows that it must be protected against the effects of continuous immersion in water.

The index of protection codes is used throughout BS 7671 to identify the minimum level of protection permissible (see Figure 5.1).

APPENDIX 6

This section provides information on the model forms required for certification and reporting of electrical installations. The model forms shown in BS 7671 are the basic format which is sufficient for compliance with BS 7671.

The forms which are supplied by a registration body such as the NICEIC or NAPIT are quite different from those in BS 7671. As an example, for compliance with BS 7671 when a new installation is completed an initial verification is carried out and an electrical installation certificate is completed, which must be accompanied by a schedule of test results and inspections. The forms which are required by most registration bodies usually have all three forms on just one certificate; for this reason the forms shown in this section are those which are supplied by the NICEIC.

APPENDIX 7

Harmonised cable core colours

Although all new installations will be carried out using harmonised colour cores, there are numerous installations which still have the old colours of red and black for single-phase installations and red, yellow, blue and black for polyphase installations. It is important when altering or adding to these installations that the conductors can be readily identified by anyone who is working on the installation. Tables 7A to 7E in BS 7671 provide information for colour coding, although the conductors can be identified by any means provided it is idiot proof.

Where we have to add to an existing installation, the colours should be matched as shown in Table A2.

In some instances, particularly in control wiring where three-core flat twin and earth PVC is used, it is tempting to connect the black of the harmonised cable to the black of the old cable. The new three-core cable will have brown, black and grey cores and the old cable will have red, yellow and blue, and the correct colour pairing would be:

- Brown to Red
- Black to Yellow
- Grey to Blue.

TABLE A2 Cable core colours

Old colour	New colour
Red	Brown
Yellow	Black
Blue	Grey
Black	Blue
Green and yellow	Green and yellow

APPENDIX 8

Current carrying capacity and voltage drop for bus bar trunking and powertrack systems

Where bus bar trunking and powertrack systems are to be used, the best method by far to ensure compliance with BS 7671 is to consult the product manufacturer, who will provide design and installation details on request.

APPENDIX 9

Multiple source d.c. and other systems

This is a specialist area which most electricians will never be involved with.

APPENDIX 10

Protection of conductors in parallel against overcurrent

It is often useful to install conductors in parallel; the most common occurrence is when we install ring final circuits. By using two conductors which are joined at the far end, our circuit can provide socket outlets over a greater area; this is due to the fact that the current can be shared between two conductors. This will result in a lower earth fault loop impedance and voltage drop.

As an example let us look at a circuit which is wired in $2.5\,mm^2/1.5\,mm^2$ 70°C thermoplastic cable supplying 13 A socket outlets.

If this circuit is a radial, it would have to be protected by a 20-A device; the maximum length of the circuit will be limited by voltage drop which is limited to a maximum of 5% (11.5 V).

To calculate the maximum length of cable we need to transpose the formula which we normally use for voltage drop:

$$\frac{mV \times A \times l}{1000} = \frac{v}{d}$$

To transpose this for length:

$$\frac{vd \times 1000}{A \times mV} = l$$

If we now replace the symbols with figures:

$$\frac{11.5 \times 1000}{20 \times 18} = 31.94\,\text{m}$$

(*The millivolt per ampere per metre value can be found in appendix* 4 *table* 4D5B.)

The maximum length of cable permitted for a radial circuit would be 31.94 m; this would severely limit the area which could be provided with socket outlets and may even result in two circuits being used.

If we were to use two conductors in parallel in the form of a ring a far greater area could be served. As an example let us use a standard ring protected by a 32-A device wired in 2.5 mm^2/1.5 mm^2 70°C thermoplastic twin and earth cable. Because the cable is in parallel, we are effectively halving the length and doubling the cross-sectional area.

This will result in the cable having a quarter of the tabulated voltage drop: 18/4 = 4.5 mv/A/m.

Now we can calculate the actual length of the cable which we can use for a ring for compliance with the voltage drop requirements.

If we assume that the circuit is to be fully loaded to 32 A, the calculation is as follows:

$$\frac{11.5 \times 1000}{4.5 \times 32} = 79.86\,\text{m}$$

For this purpose, the use of conductors in the form of a ring is extremely beneficial. If we needed to put socket outlets around a 10 m × 10 m room which was situated 15 m from the consumer unit we could possibly achieve it with two radials; this would result in using two ways in the consumer unit and two protective devices (Figure A3).

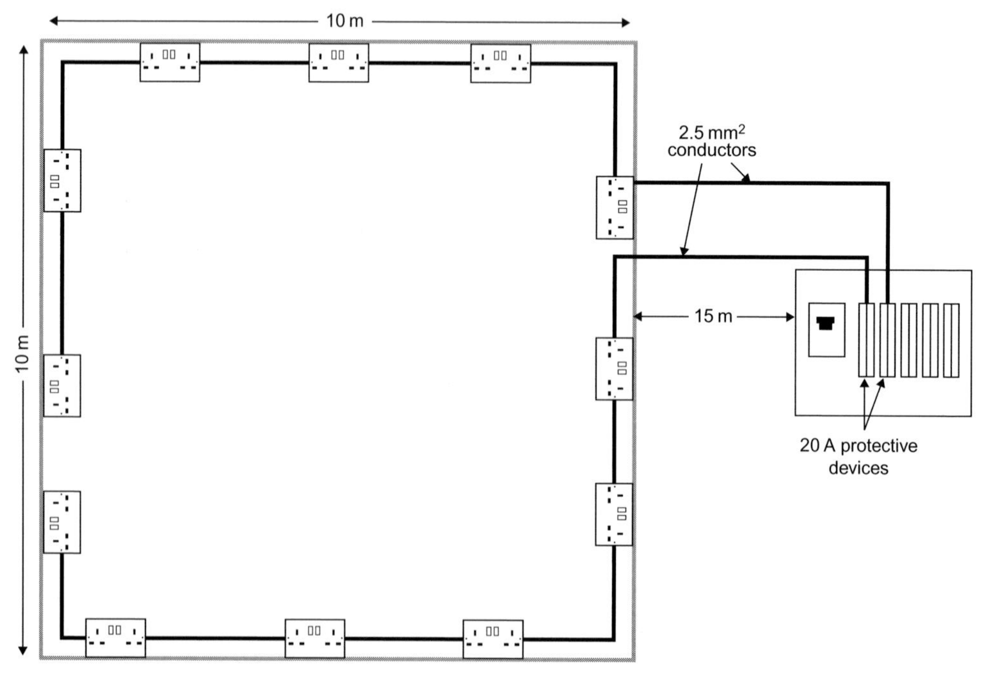

FIGURE A3 Two radial circuits

If we were to use a ring, we could achieve the same result but with only one way in the consumer unit being used and one protective device installed. This will result in a cost saving as well (Figure A4).

The same logic can be applied to larger cables used in commercial or industrial installations; it also makes the installation of cables easier as smaller cables are easier to bend than large ones. The use of conductors in parallel also often results in less copper being used to achieve the same current rating.

Let us look at table 4D4A for an example. A 50-mm^2 three- or four-core cable clipped direct (method C) from column 3 has a tabulated value (I_t) of 151 A. If we were to install two of these cables in parallel, we would have a current capacity of 302 A; to get anywhere near this value using a single cable installed using the same reference method we would have to use a cable with a cross-sectional area of 150 mm^2. Another huge cost saving.

It is important to ensure that the load is shared equally between each of the conductors; generally where multicore cables are used this is not a problem provided care is taken to ensure that the cables are installed neatly.

A greater problem with equal current sharing between conductors occurs where singles are used; in these situations advice from cable manufacturers should be sought as to the best arrangement of the conductors.

Overload and short circuit protection must be provided for each conductor in parallel. Where more than two conductors are used multiple fault paths can occur due to the conductors being connected together at each end. This could present a real

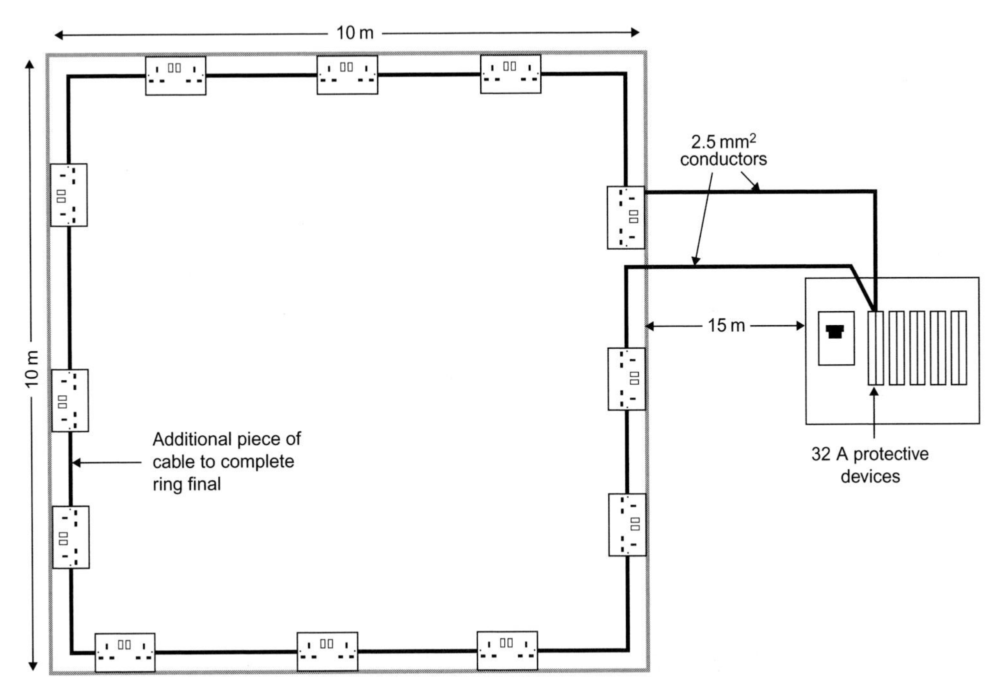

FIGURE A4 One ring circuit

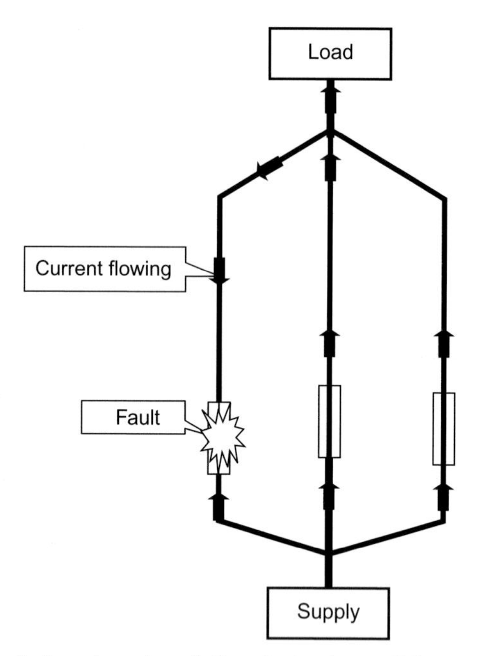

FIGURE A5 **Three protective devices at supply end of conductors in parallel**

danger as if there were to be a fault on one conductor, a current would still continue to be present on the conductor at the downstream side of the fault (see Figure A5).

One method of protecting against this would be to install a protective device at each end of the individual conductor (see Figure 7.14, Figure A6).

Unfortunately, this method of protection for conductors in parallel is not fool proof as it relies on both protective devices in the parallel conductor operating; under some circumstances this may not happen; for instance if the fault were to clear by burning open and only one device had operated one end of the conductor would still be live. A far better method of protection for this type of arrangement would be for the conductors to be protected by linked protective devices (see Figures A1 and A7).

APPENDIX 11

Effect of harmonic currents on balanced three-phase systems

It must be said that harmonics are a bit of a mystery and the way they are dealt with now may well change in the future; 30 years ago little consideration was given to them.

Due to the introduction of switch mode power supplies and other electronic devices, harmonics have become a problem in some installations; however, they are not fully understood.

Harmonics are caused by electronic equipment which converts 50 Hz a.c. to d.c. and then back to a.c. at a different frequency.

Electronic equipment does not usually have constant impedance; this is because the impedance changes due to the constant switching of the electronic components within the equipment. This switching on and off during part of the waveform results in currents being introduced back into the distribution system; this is because the

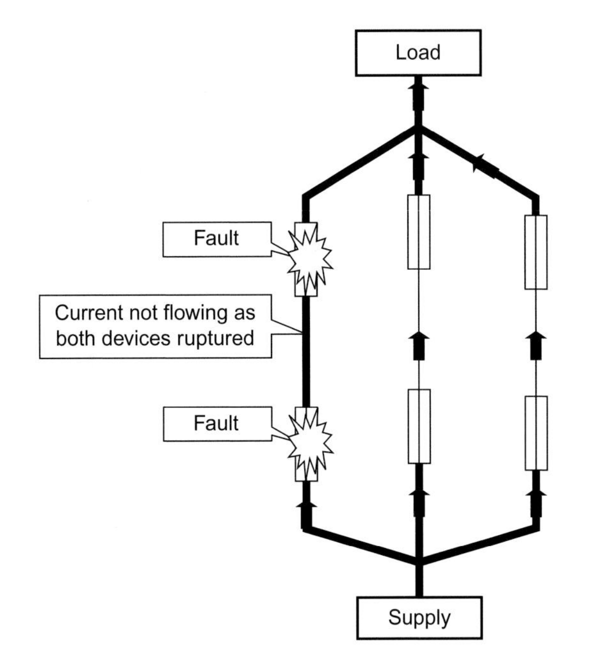

FIGURE A6 **Conductors in parallel protected at supply and load ends**

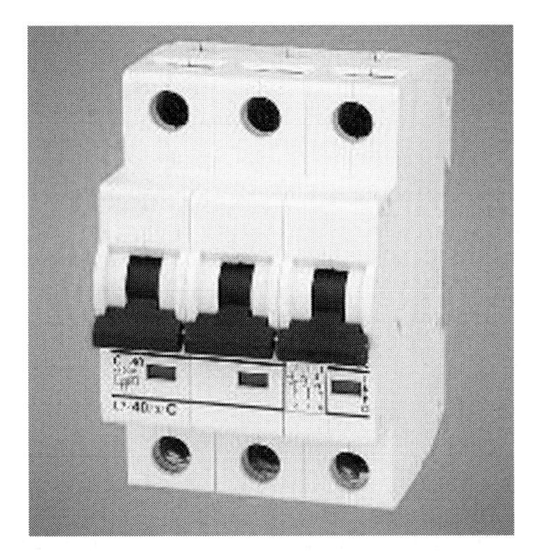

FIGURE A7 **Linked circuit breaker**

currents are at frequencies other than 50 Hz. It is assumed that a third harmonic is at 150 Hz; the effect of this is increased current in the neutral, which in certain circumstances will be higher than the current in the line conductors.

Harmonic currents are likely to be present where the following equipment is installed:

- Variable speed drives (VSDs)
- IT equipment, personal computers (PCs)
- Arc equipment
- Battery chargers

- Computer power units (CPUs)
- Electronic ballasts
- Rectifiers
- Uninterrupted power supplies (UPS)
- Discharge lighting (fluorescent, mercury, sodium, etc.).

Where this type of equipment is installed on a three-phase supply and the conductors of the circuits are fully loaded, third harmonics can cause the neutral of the circuit to become overloaded; in these instances the manufacturers of the equipment should be consulted as to the level of harmonic distortion which is likely to be caused by their equipment.

A calculation can then be carried out to ensure that the correct rating of conductor is selected.

Example 1

A balanced three-phase load is to be installed which has a design current of 26A. A manufacturer has stated that the equipment being installed has a harmonic distortion of 30%.

The rating factor of 0.86 from table 11A in BS 7671 can be applied to calculate the minimum permissible rating for the conductor.

$$\frac{26}{0.86} = 30.23\text{A}$$

The rating factor of 0.86 can be used only for equipment with harmonic distortion of up to 33%; above that the calculation has to include the percentage distortion.

Example 2

The harmonic distortion for the equipment in Example 1 is 45%. The calculation is now:

$$26 \times \frac{45\%}{100} \times 3 = 35.14$$

(3 is because it is a three-phase supply).

In three-phase circuits, the triplen harmonic neutral currents (3rd, 9th, 15th, etc.) cause the largest concern as they add instead of cancelling, since they are multiples of three times the fundamental power frequency and are spaced apart by 120 electrical degrees. Based on the fundamental frequency, triplen harmonic currents of each phase are in phase with each other, and so add in the neutral circuit. Under the worst case conditions, the neutral current can be 1.73 times the phase current.

A great deal of electronic equipment requires what is known as clean power and clean earth. Unfortunately, it is the high-frequency harmonic currents due to the switch mode power supplies which cause the distortions in the first place. These distortions can also have an adverse effect on the earthing systems of installations, in particular where there are a lot of PCs in operation. These distortions are often the cause of noise bars on PC monitors; in these cases consideration must be given to providing a copper earth mat which is connected to a very good earth and connecting each piece of IT equipment separately.

APPENDIX 12

Voltage drop in consumers' installations

This has been covered in previous parts of the book.

APPENDIX 13

Measuring the insulation resistance or impedance of floors and walls to earth or the protective conductor system

This section describes how to measure the impedance of a room which is going to be used as a non-conducting location. The description of how to carry out these tests is given in appendix 13. As this type of protection is rarely if ever used in the UK no further information is required.

APPENDIX 14

Measurement of earth fault loop impedance taking into consideration conductor resistance increase due to temperature increase

This has already been covered in earlier chapters but it does seem to cause some confusion and for that reason it is worth going over again.

When the Z_s of a circuit is measured the temperature of the conductors is not known; also, the Z_s value will include the value of Z_e as well as R_1 and R_2.

As an example let us take a circuit protected by a 20-A BS 1361 fuse, and wired in $2.5\,mm^2/1.5\,mm^2$ twin and earth cable which is 32 m long. The Z_e for the supply is $0.8\,\Omega$.

If we look in table 41.2, we can see that to ensure operation of the fuse within the permitted time, the maximum Z_s permissible is $1.7\,\Omega$.

The measured value of Z_s at the furthest point of the circuit is $1.4\,\Omega$. Unfortunately, we cannot compare this measured value to the maximum permitted value because we do not know what the actual temperature of the conductors is.

Appendix 14 allows us to carry out a rough calculation which will take into account any rise in temperature due to current which may flow in the circuit; this calculation is known as the 'rule of thumb'.

We must multiply the maximum value of Z_s by 0.8.

$$1.7 \times 0.8 = 1.36\,\Omega$$

This value must now be compared to the measured value, which was $1.4\,\Omega$.

If this value (1.4) is less than our recalculated Z_s (1.36), the circuit will be fine and the protective device will operate as required.

In this example the circuit will not be suitable as the measured value (1.4) is greater than the recalculated value (1.36).

This does not necessarily mean that the circuit is completely unacceptable; the calculation is only a rough guide to prove compliance and the values are always on the right side of good. In situations where the values are close, a more accurate calculation can be carried out as it may prove that the circuit is satisfactory after all.

We first need to obtain the value of $R_1 + R_2$ by measurement or calculation. In this case, we can calculate it as we know the length of the circuit is 32 m.

The value of $r_1 + r_2$ given in table 9A of the on-site guide is 19.51 mΩ/m at 20°C. The resistance of 32 m of this cable must be:

$$\frac{19.51 \times 32}{1000} = 0.624\,\Omega$$

But of course this is at 20°C and we know our cable may reach 70°C when it is carrying current. This rise in temperature is calculated by using the figures given in table 9C of the on-site guide; as this is a 70°C cable the value we are using to compensate for the rise in temperature is 1.20.

We must now multiply the resistance of our cable ($R_1 + R_2$) by 1.2.

$$0.624 \times 12 = 0.748 \text{ (rounded up to 0.79)}$$

The accurate Z_s value of our circuit can now be calculated.

$$Z_s = Z_e + R_1 + R_2$$

$$Z_s = 0.8 + 0.748$$

$$Z_s = 1.54\,\Omega$$

If we now compare this value with the maximum permitted value of Z_s, which is 1.7 Ω, we can see that it is less, which means that the measured value of the circuit is acceptable and that the rule of thumb calculation really is just a rough guide. For any circuit which is close but not acceptable it is worthwhile carrying out a more accurate calculation.

APPENDIX 15

Ring and radial final circuit arrangements

Socket outlet circuits have often caused confusion, particularly where a reduction in cable size may be required. The diagrams in this appendix make what is permitted very clear (Figure A8).

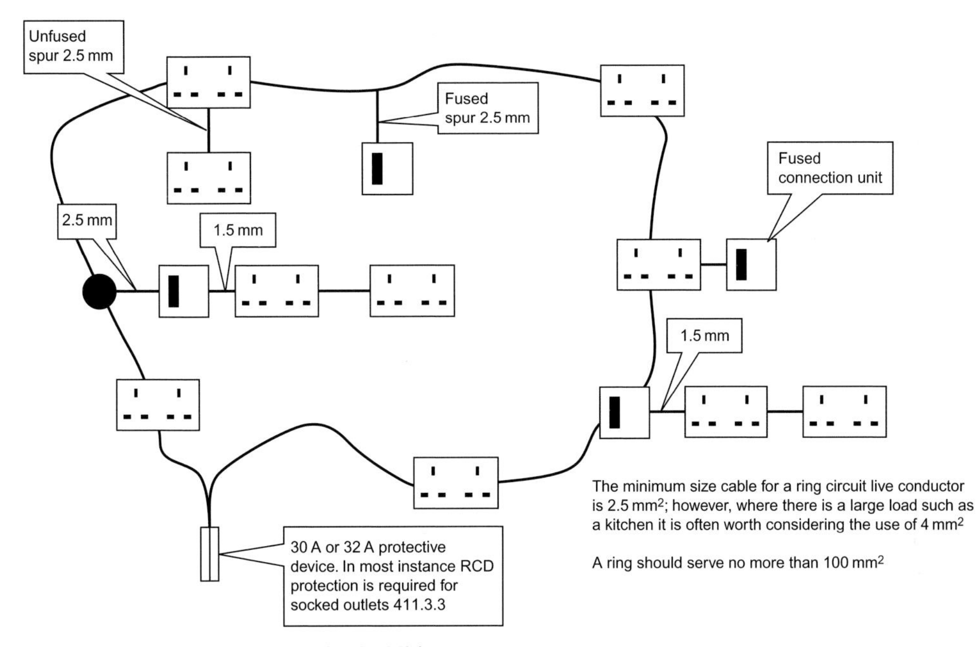

FIGURE A8 Ring circuit with permitted additions

Index

P

R

S

T